the WORLD According to the NEWEST and MOST EXACT OBSERVATIONS

the NEWEST

MAPPING ART + SCIENCE

the WORLD *According to* *and* MOST EXACT *OBSERVATIONS*

SUSAN BENDER *and* IAN BERRY *with*
BERNARD POSSIDENTE *and* RICHARD WILKINSON

The Tang Teaching Museum and Art Gallery at Skidmore College

This catalogue accompanies the exhibition

The World According to the Newest and Most Exact Observations: Mapping Art and Science

March 3 – June 3, 2001

The catalogue is made possible through support from the Henry Luce Foundation and the Friends of Tang. The exhibition has been funded in part by MapInfo Corporation and the Friends of Tang.

This catalogue is dedicated to all of those whose vision for the Tang Museum has made this possible, especially David Porter and Phyllis Roth.

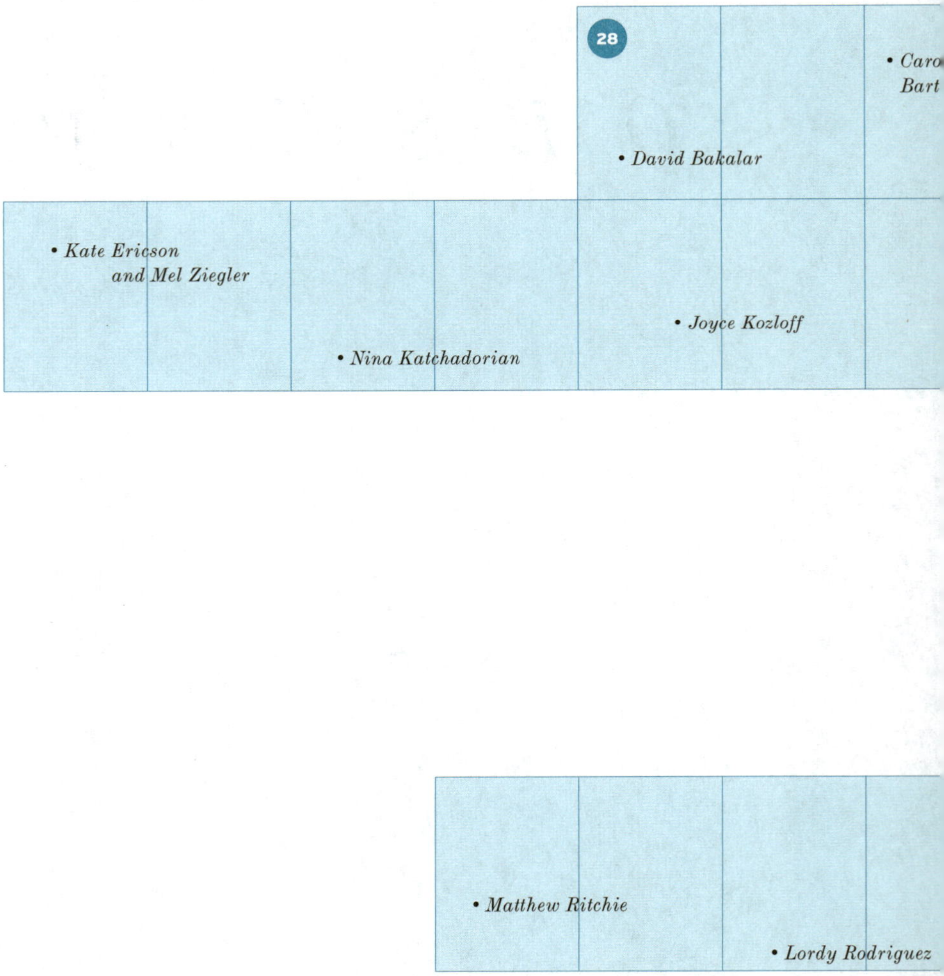

INTRODUCTION

Every map is a cultural construction that geographers, scientists, and artists alike create to make and convey meaning. This catalogue documents an exhibition that focuses on the human body and the political geography of northeastern North America through maps, atlases, and scientific models from the 16th century to the present. These are combined with the work of eighteen contemporary artists whose installations, paintings, photographs and sculpture expand notions of how and why we map. Each of these objects captures for us the world view of its maker — and allows us to ask how that point of view imbeds itself in the map, as the mapmaker decides how to render a landscape, anatomy, molecule, or idea into a simplified model of a complex reality.

The greater the discrepancy between the size of the feature being mapped and its re-presentation, the greater the potential for its distortion. In the case of geographic maps, the distortion occurs as a round surface is projected onto a flat plane and a large area is compressed into a small space, leaving little room for the variety of details that could be included. In the case of maps of the human body the modeled systems range in size from an entire body to microscopic views that encompass a variety of functions at each level. It is in the process of deciding what projection to use and what features to include that the mapmakers reveal to us their worldview. In these sequential decisions of emphasis by way of color, size, and selection of detail, mapmakers, anatomical illustrators, geneticists, and in some cases artists, convey their sense of what is important, an estimation that is deeply imbedded in their cultural location.

From this perspective, then, we can approach every map as a complex visual text that represents the world through the culturally modified lens of its maker. As we navigate the many layers of meaning encoded in the map,s words and images, we are reading our way into the mapmaker,s world. We encounter this world by asking ourselves such questions as why has the mapmaker chosen this title and these place names? What features have been selected for presentation in word and image? What message does this convey about ownership? What has the mapmaker not included in the map and what do the empty spaces represent or imply? How does the perspective of projection chosen by the mapmaker emphasize certain features of the reality being mapped? The answers to these questions are not encoded in the subjects, but are instead attributed to the subjects by the scientist, geographer, or artist who authors the map.

The western mapping tradition illustrated in this exhibition developed in concert with the Renaissance and the Age of Exploration, and it laid the basis for a cultural tradition that Denis Wood has described as "map saturated." By this, he means that we map everything routinely and that we use maps as everyday guides to our world. From the network of computer connections that deliver information to deeper understandings of our molecular structure, the act of mapping is front-page news. Thus for any part of our fast changing world we can find innumerable maps that attempt to describe it. We can use these maps to enter the world as their makers understand it and in that process transform our own worldview.

SUSAN BENDER and IAN BERRY

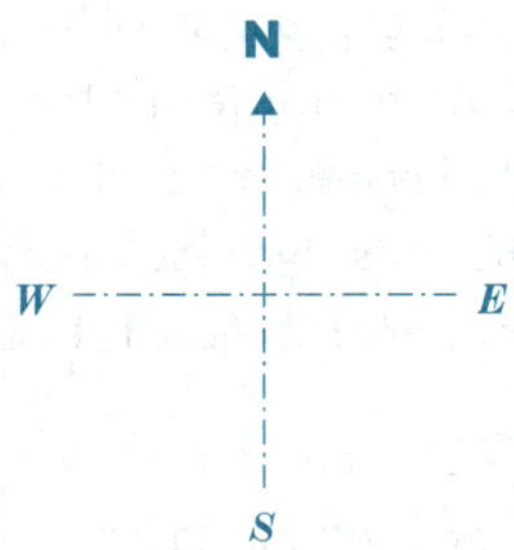
N
W
E
S

MAPPING ART + SCIENCE
SELF
RTRAIT AS A
PROPORTION
N THIS BLACK
ENTING LIFE L
REMAINING
PACE REPRES
COME, BASED
CAL LIFE EXP

HERMAN MOLL
New and Exact map of the Dominions of the King of Great Britain on Ye Continent of North America
1735

Seeing the World in Maps

SUSAN J. BENDER

TIME AND SCALE

THE MAPS ASSEMBLED IN THIS EXHIBIT SEEK TO RECOVER SUCCESSIVE MAPPINGS OF SARATOGA Springs, New York, the cultural locus of its genesis. The earliest maps included date to the late 16th through the mid-17th centuries and describe European explorers' and settlers' encounters with coastal northeastern North America as well as their initial penetration of the Hudson River valley as far north as Albany (then named Fort Orange). These small-scale maps describe large tracts of land, and are highly schematized – revealing much about their makers' world view. While Saratoga Springs does not appear on any of these maps (since that place was not yet known to the colonizing powers) these maps can help us to understand the world into which this town and city would subsequently be introduced.

The 18th century maps document the process of colonial domination of this area. They range from relatively large-scale presentations of a delimited territory in the upper Hudson drainage to small-scale maps of the entire eastern seaboard of North America. While maps of the former type tend to express patterns of colonization through the expansion of small settlements, the latter types present to their audiences much larger sociopolitical entities such as "the Dominions of the King of Great Britain" or "the British Empire in America." While the place name "Saratoga" appears on these maps as early as 1733, its referent is Philip Schuyler's settlement later to be named "Schuylerville." Though known to European traders and settlers by the end of the 18th century, Saratoga Springs had yet to achieve the economic or political status that would warrant its appearance on maps of such scale.

OPPOSITE
L. R. BURLEIGH
Saratoga Springs, NY
1888

RIGHT
VENICE FRISIUS/APIAN
Chartes Cosmographiques avec les Noms Proprietez, Nature & Operations des Vents
1581

BELOW
HONDIUS
Nova Anglia Novum Belgium et Virginia
1638

The 19th and 20th century maps in this exhibit focus increasingly on Saratoga Springs and its environs and are much larger scale projections than those from earlier periods. At this scale, the mapmaker can choose to project a great deal of detail about this place – much as *DENIS WOOD* has done for his Raleigh, North Carolina neighborhood (pg. 80-81). Many of the maps of Saratoga Springs show us the town and city developing in concert with each other and in response to the tourist industry that increasingly helped to shape the region's identity. The earliest of these representations dates to 1810 and depicts *GIDEON PUTNAM*'s initial claims on and development of the land surrounding Congress Springs, the area that was to become the city's center (pg. 19). By 1888, *L.R. BURLEIGH* mapped out a "bird's eye view" of a densely populated resort town, replete with race track, baths, springs, hotels, and other forms of entertainment (pg. 13). Still other maps, dating up through the mid-20th century continue to project information about the city's "points of interest" to its visiting guests. Throughout this same period, we can also find a local mapping tradition that is aimed at the interests of the area's residents. Here the discourse is about folklore, ownership, and regional travel. They tell us what the people who lived here were talking to each other about.

Thus both time and scale help to frame every map's message in some fundamental way. What the mapmaker can say about a tract of land depends initially on these two factors. The smaller the scale of a map, the less detail and more generalized the presentation. But it is precisely in such maps that the selective lens of the mapmaker can be overlooked. They lack the "local color" of the larger scale maps. But, every map is a model of a spatial phenomenon that exists at a much larger scale, and the

smallest scale maps involve the greatest selection. We should not be mislead by their abstract and generalized quality; we need to look for what it is the mapmaker wants to tell us in this constructed image. And, what the mapmaker wants to tell us is based on his moment in time. Time influences both the basic empirical data upon which the map is based (i.e., what we know about an area), as well as the map's worldview of a place (e.g., as wilderness or hometown). If we want truly to read a map, we must know how time and scale are structuring the presentation of place.

ACCURACY

IT IS UNUSUAL, HOWEVER, FOR US TO BEGIN INTERROGATING A MAP FROM THIS PERSPECTIVE. WE tend instead to ask whether the map is accurate and then allow ourselves to be annoyed, amused or captivated by a map's inaccuracies — depending upon our immediate purpose in viewing it. In general, the older the map the more we are amused by the seemingly naïve construction; the more we are interested in "how much they didn't know then." But, through this exhibit we urge the viewer to abandon this presentist perspective and accept instead that these maps were created according to "The Newest and Most Exact Observations" of their time. It doesn't take us long to locate all the inaccuracies imbedded in *HERMAN MOLL*'s 1735 map entitled ***A New and Exact Map of the Dominions of the***

King of Great Britain on Ye Continent of North America (pg. 10). What is interesting about this map is, however, not that it is inaccurate according to early 21st century standards, but that it is in fact a new and exact map for its time. It is then an accurate cultural document that tells us not only about the state of English geographical knowledge of northeast North America in the early 18th century but also about what this mapmaker thought were the most important images to associate with this landscape.

We can and should marvel at the accuracy with which the coastline and St. Lawrence drainage are mapped. Indeed the exhibit includes several maps that project the familiar outline of the eastern seaboard relatively accurately as early as the mid-17th century, even as interior regions were relatively poorly understood. What the latter tell us is that the early seaborn Dutch, English, and French explorers mapped out the details of North America's east coast almost immediately, and together with the detailed descriptions of important landmarks that accompanied these maps in atlases – the common publication format for maps in the 17th century – they created a text that enabled others to follow in their footsteps. *JACOB ROBYN*'s 1689 ***Sea Lightning Atlas*** is a marvelous example of this type of publication. In it, we can locate a wonderfully precise mapping of the New York harbor area, including sea depths, and this map is accompanied by minute textual descriptions of landmarks and place names.

But it is also easy for us to pick out the many inaccuracies in these mappings. A quick appraisal of *MOLL*'s map, for example, reveals many mistakes with regard to the placement and dimensions of inland bodies of

ABOVE LEFT
JAN JANSSON
American Septentrionalis Amstelodami
1635

ABOVE CENTER
VISSCHER
Novii Belgii novaeque Angliae nec non partis Virginiae Tabula....
1685

ABOVE RIGHT
LOTTER/SEUTTER
Recens edita totius Novi Belgii in America Septentrionali...
1757

RIGHT
ARNOLD COLOM
Pascaarte van Nieu Nederlandt uytggeven door Hendrick Doncker
1656

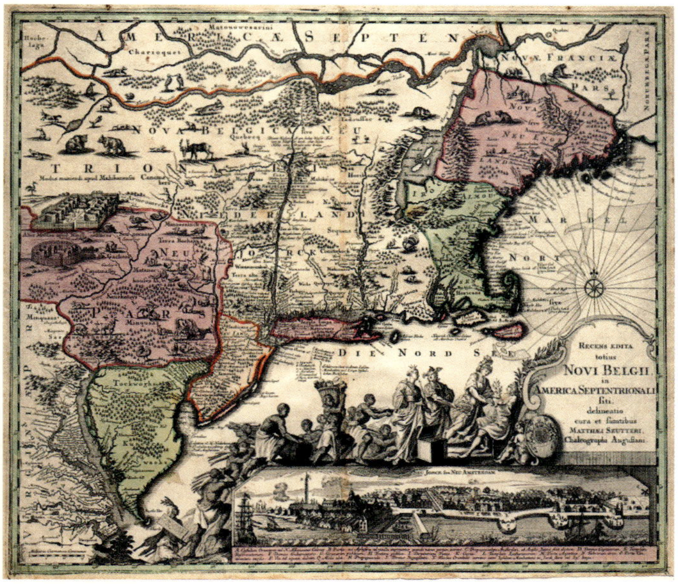
RECENS EDITA
totius
NOVI BELGII
in
AMERICA SEPTENTRIONALI
siti.
delineatio
cura et sumtibus
MATTHÆI SEUTTERI.
Chalcographi Augustani.
DIE NORD SEE
MAR DEL
NORT

Pascaarte
van
NIEU NEDERLANDT
uytgegeven
door
Hendrick Doncker.
t'Amsterdam
NIEU NEDERLANT
Nieu Engelant
Virginia
Minquaes
Makima
Sequins
Pequatoos
Siwanoys
Langh Eylant
alias Matouwacs
Sanhikans
Matovancons
Aquauachuques
Ermomex
Nieuw Amsterdam
Groote Rivier

LAKE HURONS
Baye de Toronto
Algonquins
ILLINOIS
LAKE ONTARIO or FRONTENAC
LAKE ERIE
PENNSYLVANIA
Nation du Chat
destroyed by the Iroquois
APALACHE
MOUNTAINS
NEW YORK
NEW
JERSEY
Montreal I.
Miamis
Onondages
LAND

Cherokees
Chaovanon
SOUTH
CAROLINA
FLORIDA
Apalatche B
Cape Charles
Cape Henry
Currituck Inlet
Roanoke Inlet
Cape Hateras
C. Look out
Sandy Bay
Cape Fear
Cape Carteret
St Augustine
Matanzes I.
Mosquitos I.
Isle d Anclote
The Southern Bounds of CAROLINA, by the 2d Charter

OPPOSITE
HENRY POPPLE
A Map of the British Empire in America with the French & Spanish Settlements adjacent thereto
1733

ABOVE
Map of the Northern Parts of New York
1758

RIGHT
The Hudson River from New York to Albany
18th century
Scrimshawed powder horn

water and other geographical features. Yet these inaccuracies carry valuable information for us. They describe the routes of colonization (by what was unknown as opposed to known), and detail the importance of waterways to this venture. Similarly, the inaccuracies of the smaller scale maps tell us that the far northwest coast of North America was the last part of the western hemisphere to be penetrated and mapped by the colonizing powers. At the same time, *MOLL*'s claim to have presented us with "A New and Exact Map" seems hopelessly naïve. We want to ask, how could he make such claims? Did he not know how much he did not know? Quite possibly, from his cultural location, he didn't understand the scope and variety of information precluded in his mapping. *MOLL* was in fact presenting his audience with a map based on "the newest and most exact observations."

What we propose in this exhibit is that mapmakers can routinely make such claims. And we suggest that mapmakers of the early 21st century in the United States are no less immune to the misleading assumptions informing such claims than was *MOLL*. To be sure, the observations informing contemporary map projections are far more precise and comprehensive than those available to *MOLL*, but does the presentation have greater truth claims? *WENYON + GAMBLE*'s photograph of the Mt. Haystack Observatory telescope's mapping of itself raises this question, as well, only through image instead of text (pg. 76-77). Recall that we could marvel at the accuracy of the 17th and 18th century maps as we saw the familiar shape of North America's east coast appear in them. But is that familiarity, based on a projection developed by Mercator in 1569, necessarily accurate? Could it, too, be a culturally agreed upon construction? Why is Mercator's projection, one that favors maintaining accuracy of shape over size, the one that we continue to use? Why have we not abandoned this projection for that developed and published by Peters in the 1990's? Even as Peters accepts some distortion of shape to create accuracy in relative size, we are uncomfortable with the image thus presented. North

LEFT
LE SR. ROBERT DE VAUGONDY, PARIS
Partie di L'Amerique Septentrionale, qui comprehend le de L'Ohio...
1757

OPPOSITE
JAMES SCOTT
Property of Gideon Putnam
1810

American and Europe appear oddly misshapen and certainly too small in the Peters' projection. Or, as *JOYCE KOZLOFF* suggests with her globe that depicts Islamic mapping traditions (pg. 42), why does north have to be up? We cannot escape the conclusion that no matter how accurate the observations informing a map, its claim to exact replication is as open to interrogation now as it was in the past.

ECONOMICS OF MAPPING

THUS WE CAN RETURN TO MOLL'S MAP AND ASK, "IF ITS ACCURACY RESIDES IN A DESCRIPTION OF how North America was penetrated by colonizing powers, might it also tell us something about why this process occurred?" Indeed it can, together with several other maps included in the exhibit. Here the map's imagery becomes a central concern, for this is the stuff upon which European audiences constructed their emerging vision of the what this land was about. And, *MOLL*'s use of the 18th century convention of the "beaver inset" is most instructive in this regard. In it, we encounter a view of the wondrous riches of the New World. A vision of a veritable storehouse of goods in high demand, but short supply in England emerges from this image (see *A. J. CRONIN*'s ***Changes in the Land*** for a marvelous development of this idea). The seemingly endless, detailed lists of North America's natural goods found in *JOHN SMITH*'s 1632 ***The Generall Histories of Virginia, New England, and the Summer Isles***, confirm that this perspective informed the very earliest European activity on the continent. The images surrounding the inset of *LOTTER AND SEUTTER*'s 1757 Map (pg. 15), where slaves carry the products of the New World to a European ruler tell us about the durability of that perspective – as does *MOLL*'s presentation. Finally, *POPPLE*'s 1733 mapping of ***...the British Empire in America with the French & Spanish Settlements adjacent thereto***, reveals through the use of

LEFT
L.H. CRAMER & J.W. MOTT
Map of the Town of Saratoga Springs
1879

OPPOSITE TOP
HENRY LAWRENCE REALTY
Map of Saratoga Springs and Excelsior Park
1881

OPPOSITE OBOTTOM
LESTER BROTHERS REALTY
Drives in the Vicinity of Saratoga Springs
1887

toponyms the competing claims of the colonizing powers to these riches (pg. 16). Even as French and Spanish claims are acknowledged through settlement names, English claims symbolized in place names are the dominant features of Popple's mapping. Nowhere is this more evident than in the notation in the upper lefthand corner of the lower map, where "A fit place for an English factory" (note: the term "factory" denotes a trading outpost in the language of the day) is written into a relatively unclaimed area (noted through native toponyms) bordering Spanish-named settlements. These maps and accompanying texts tell us as much about the colonization process and its contested claims as any scholarly essay.

In the same vein, the successive mappings of Saratoga Springs and it environs beginning in the mid 19th century were structured by an underlying assumption of the primacy of economic processes. In our world, economic factors condition not only how we interact with each other, but also with the landscape, and these interactions in turn influence how the landscape is mapped. Herein lies the perspective from which we may read in the series of Saratoga maps in this exhibit. For example, Saratoga Springs is mapped most regularly in relationship to Saratoga Lake. Even as *CRAMER AND MOTT*'s 1879 street map presents numerous details about Saratoga's street plan and land ownership patterns, the city itself is visually referenced in relation to Saratoga Lake. Why should this be? The 1887 *LESTER BROTHERS*' (local real estate brokers) map of ***Drives in the vicinity of Saratoga Springs*** provides us with some insight into the economics underlying this mapping convention. In the Lester map, we see a number of concentric circles used to convey information about the distances involved in travel from town. Even as this projection creates a the sense of Saratoga Springs as a central place, it is visually overpowered by the scale of Saratoga Lake and the relatively developed road systems to the south and east of town. At the same time we know from tourists guides of the era that carriages

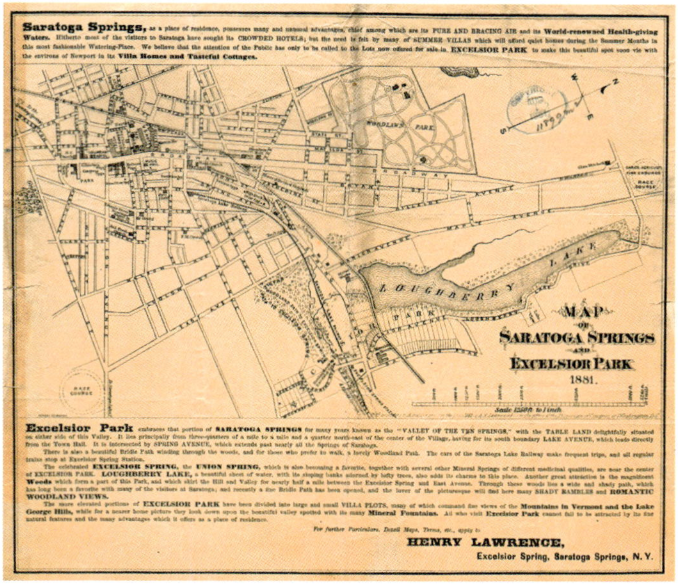

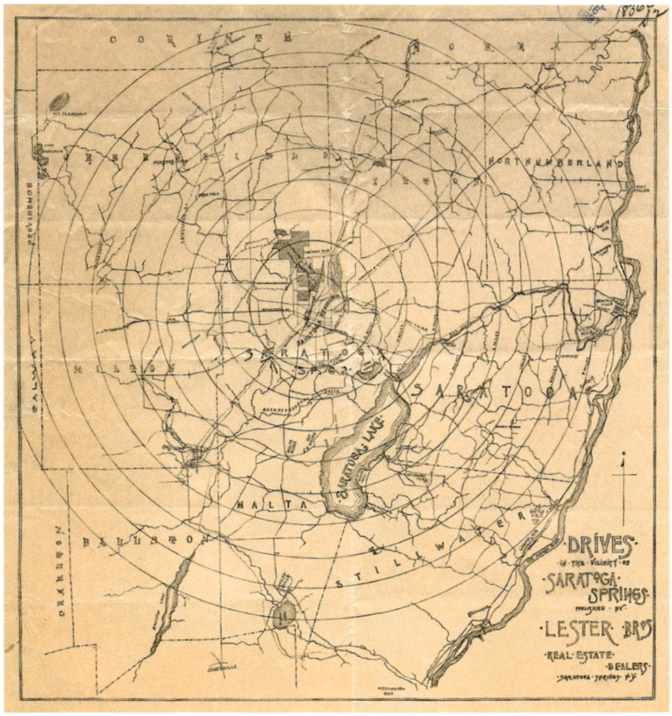

drives from town were a major amusement for tourists and that Saratoga Lake and its attractions were a frequent point of destination for these outings. As a key part of the town's tourist economy, Saratoga Lake was of necessity appropriated into Saratoga Springs' mapping.

From place name to the manipulation of images (be they decorative or parts of the geographical presentation), maps need to be read for their economic constructions, and in our culture economic power is almost always intertwined with political power.

POLITICS

UNDERLYING MUCH OF THE PRECEDING DISCUSSION IS THE commonly held assumption that a primary function for maps is to tell us about who owns what parcels of land and to delineate the boundaries among these parcels. And, since land ownership is bound up with political power, maps can provide profound insight into processes of domination — especially in the case of a land subject to colonization such as northeastern North America. Even as the titles of several the 18th maps discussed above (e.g., *MOLL*'s ***...Map of the Dominions of the King of Great Britain***

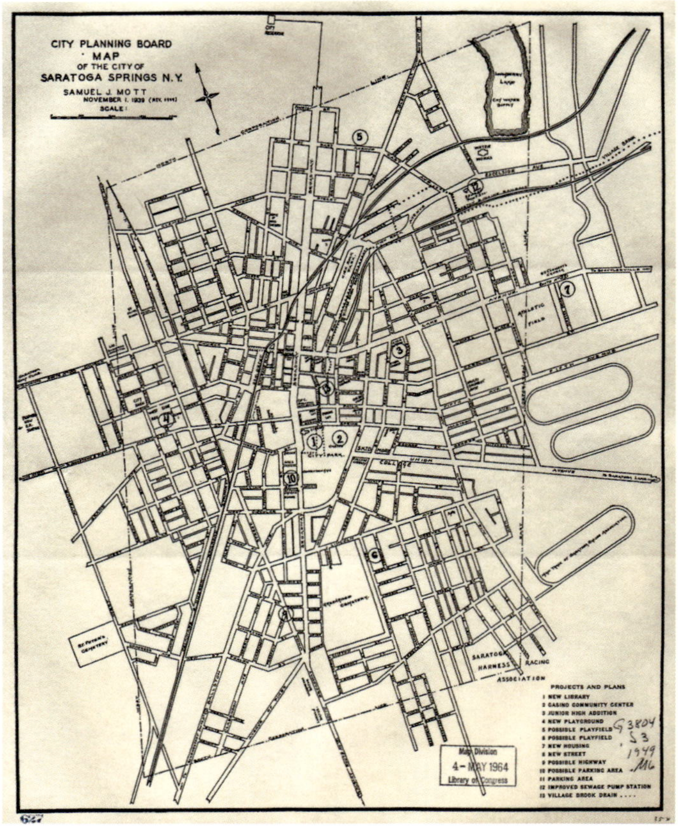

RIGHT

L.H. LAWRENCE AND SAMUEL J. MOTT

City Planning Board Map, revision of the 1939 map

1949

BELOW

INTERSTATE PUBLISHING COMPANY

Map of Saratoga Springs, New York

1955

OPPOSITE

J.A. COREY

Memory Map

1936

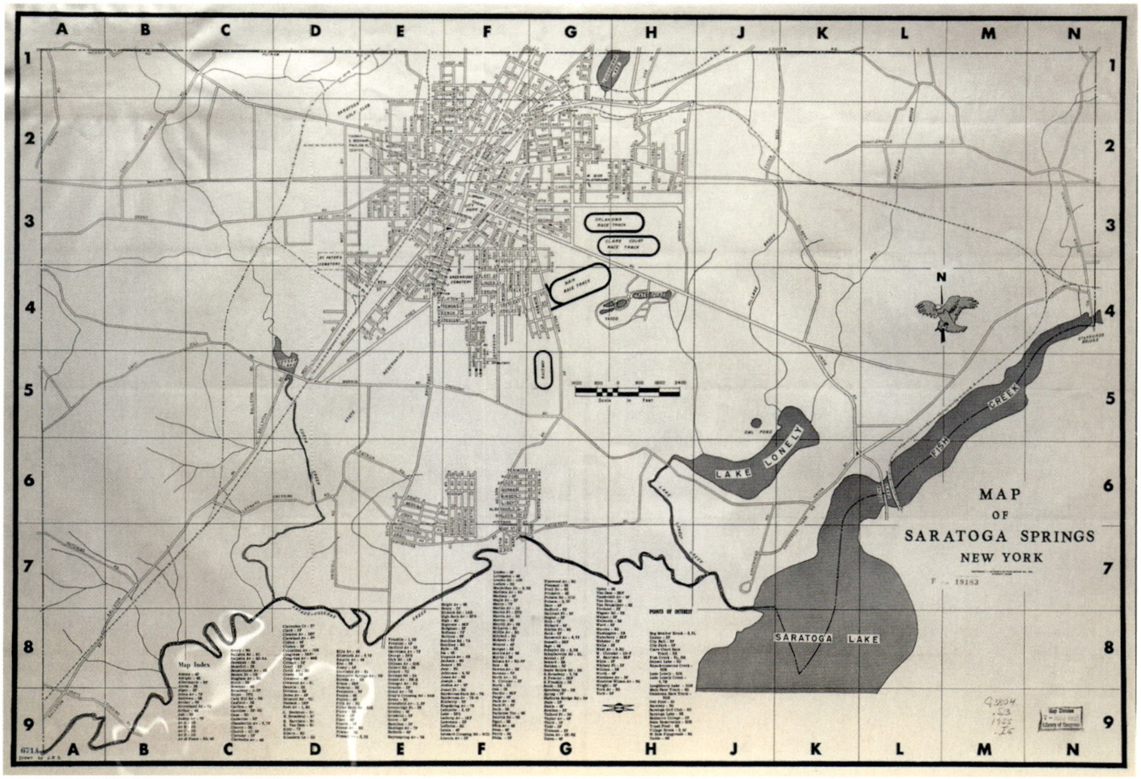

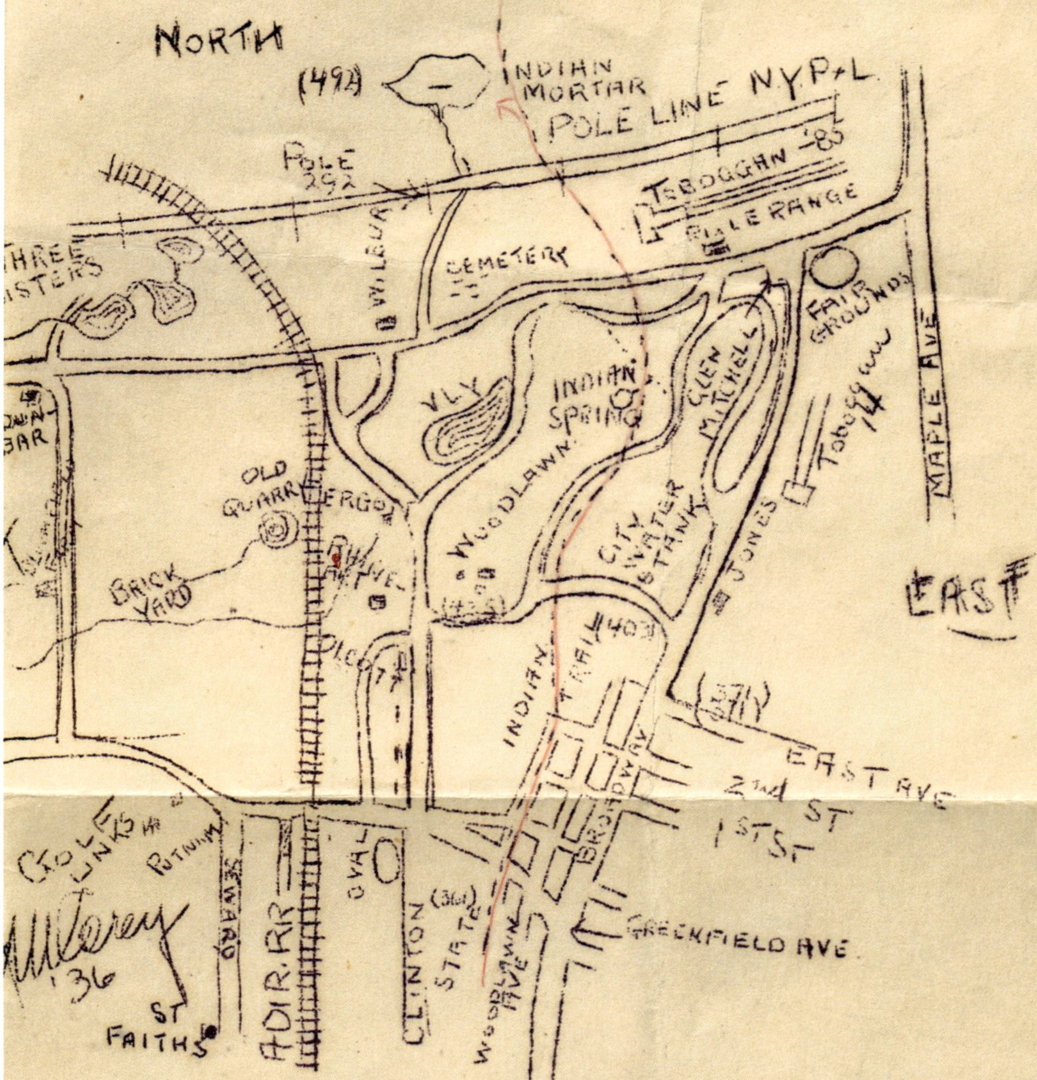

The above memory map, not drawn to scale but all points within a radius of two miles marks spots mentioned by various writers relative to Saratoga Springs but not located so that a person interested might find them. Places mentioned include the "Three Sisters", "Toboggan Slides", Indian Spring, Indian Mortar, Indian Trail, Old Brick Yard, Woodlawn Oval, Fair Grounds, Rifle Range, The Vly or Love Lake, The territory mapped is easy of access and there are many trails included that the hiker and skier will find interesting. On some of the trails markers are placed denoting that it is a ski trail. A person should not go beyond the power line (N.Y.P.& L.Corp.)unless familiar with the territory as it is covered with brush and swamps and one may easlily lose his or her way. Skiers unless experienced should not attempt the hills in the back country as there are numerous ledges, ridges and boulders that may prove dangerous. The short trail from pole #292 to the Indian Mortar (300 yds) is plainly marked and trail should not be deviated from. If you visit the mortar take Glen Mitchell or Miller Road to Wilburs, turn north at private road and follow trail. If you visit the Indian Spring turn in last Woodlawn Park gate on Broadway take the easterly road bearing to the left at foot of hill till spring is reached. Trail is marked. If on a trip to the Three Sisters go out Seward St. to Dunbar's corner, leave car and take old road north for quarter of a mile. Old toboggan slide (1885) was on first rocky ledge just north of Glen Mitchell. Second slide (1914) was just north of Jones Est. Fair Grounds Rifle Range and Glen Mitchell were at foot of hill on B'way. There are many old roads and trails within this territory that are not shown but will bring a person out on Broadway, Clinton St. or the Adirondack R.R. Thru error "The Old Cold Spring" west of the Rhinehart estate and near the Ad. R.R. was not placed on the map. A well worn path will take one to it.

on Ye Continent of North America) proclaim ownership by colonizing entities, their iconography and text can tell us much about how this process occurred. For example, a comparison of the insets from the 1685 and 1757 maps of New Belgium and New England reveals a move away from picturing the native inhabitants of these lands to picturing the slaves who became essential to these lands' economic productivity in the service of the crown. At the same time, native imagery and place names are increasingly replaced by European-derived nomenclature even as boundaries outlining colonial claims of ownership are on the increase and these political territories are more tightly defined.

The relatively large scale 1788 manuscript ***Map of the Northern Parts of New York*** – dealing as it does with the presentation of local landscapes—provides us with a valuable description of how this process occurred (pg. 17). Here we encounter the first signs of lands being divided, parceled and claimed by household units pictured as tiny houses along the various rivers that were the primary routes of transport in this era. Moreover each of these households is denoted by the "owner's" name. One can easily imagine how this mapping affected the world view of its user, describing a land whose primary features were the households of European landowners. This mapping, in other words, reveals the process of European domination at the local level, and this is the level that underpinned the grand claims made by *MOLL*, *POPPLE*, and others. And note, as well, how central the process of parceling out the ownership of land is to both levels of mapping.

An 1803 map of ***Lands Adjacent to the Fish Creek***, provides a poignant reminder of the impact of this process on Saratoga's native inhabitants. This map was accomplished as part of a project undertaken by New York's Surveyor General to map and parcel all privately unclaimed land in the State, land that was thus under control of the state government. Here we see lands that were parceled and identified either by individual landowner's name or by number for those parcels as yet unclaimed. The one

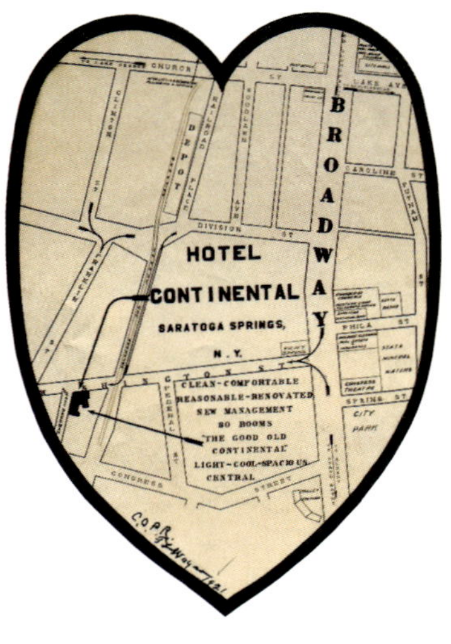

exception to this scheme is a single tract of land labeled "Indian Reservation" with a few "Indian Houses" drawn onto it. Gone are the large tracts of land signified by tribal identifier that were so prominent on mappings during the first half of the 17th century; "Mahican" was the tribal claimant in the upper Hudson area. But, by 1803 tribal identity is irrelevant to the mapmaker, and native claims to landownership are reduced to a small parcel bounded on all sides by claimants from a colonizing cultural tradition. After this point in time, native identity is erased from all maps that deal with political and economic control. Significantly, the most recent map in the exhibit that links native American identity to the local landscape is *J.M. COREY*'s ***Memory Map*** of Saratoga Springs (pg. 23). We can read this map as an "insider's" map to Saratoga Springs, drawn by a prominent community citizen to communicate with other area residents about a local landscape bound up with folkloric traditions. Here the regular features of scale, streetscapes, political boundaries and landownership are abandoned to communicate other meanings. This is a map about the stories told about this place; it is not about power.

CONCLUDING OBSERVATIONS

IN THE WAYS THAT I HAVE TRIED TO OUTLINE ABOVE, ISSUES OF SCALE, TIME AND PERSPECTIVE combine to influence the mapmaker's art and science. Each map tells us something about what the mapmaker knew and wanted us to know. It can even reveal something about how the mapmaker came to know this information. If examined from this perspective, the corpus of maps produced in our cultural tradition can tell us a good deal about how we have produced knowledge about geographic space, about the meanings that we consistently associate with space. In this exhibit we have explored a map series that reveals the most common of those associations. Mapping here is about exploration, settlement, and claims to ownership – all of which are traditionally male undertakings in our cultural tradition. Moreover, we cannot fail to recognize that knowledge production in the West has been predominantly a male undertaking, and mapping is no exception to this rule. Thus even as we have sought to examine what the maps in this exhibit are about, we should not, in closing, forget what they are not about. The cadre of male mapmakers whose work is represented herein are describing a world that they sought to know and dominate, the two being inextricably intertwined. I therefore want to end with reference to the one rather indiosyncratic voice in this exhibit. *ALFRIDA STORM*, former Skidmore faculty member, has mapped for us – in full local color – the world that she knew and navigated, Skidmore College's 1939 campus (pg. 26). A stylized treat to the eye, her map is not the stuff of the great 17th century atlases, it is rather an invitation to consider a parallel, largely unexplored mapping tradition that tells us something about social space. Hinted at by the artists' whose work is represented in this exhibit, we would do well to attempt to recover this mapping expression as well. ◉

OPPOSITE TOP
J. D. FISK
New York, Saratoga Springs (City)
1968

OPPOSITE BOTTOM
F.L. WAGAN
Hotel Continental Saratoga Springs, New York, *1921*

BELOW

ALFRIDA STORM
Skidmore College
1939

OPPOSITE

CHAMBER OF COMMERCE
Fun and Rest... Be Saratoga's Guest. The "Vacation" Convention City
1952

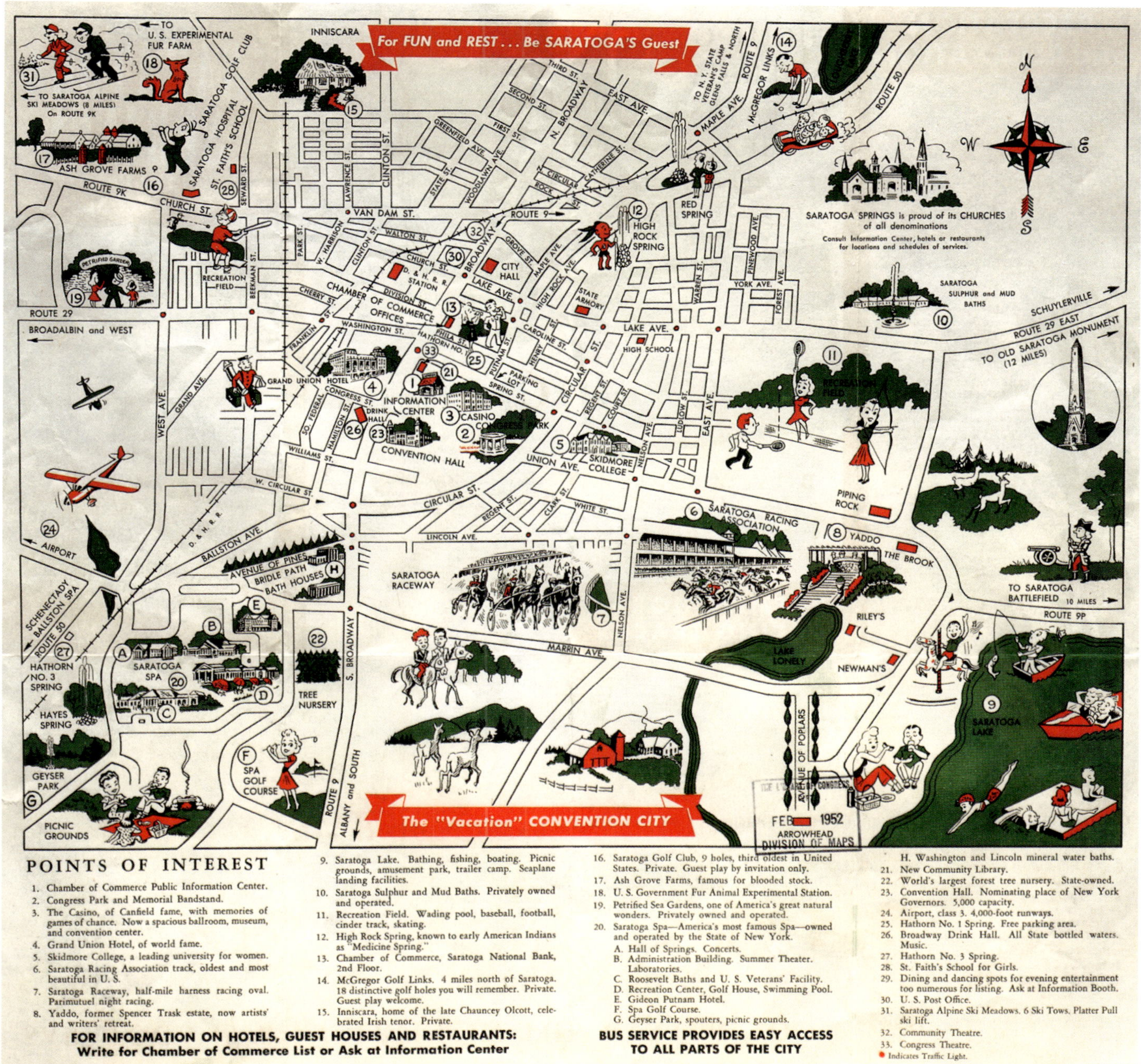
For FUN and REST . . . Be SARATOGA'S Guest
The "Vacation" CONVENTION CITY
SARATOGA SPRINGS is proud of its CHURCHES of all denominations
Consult Information Center, hotels or restaurants for locations and schedules of services.
TO U. S. EXPERIMENTAL FUR FARM
TO SARATOGA ALPINE SKI MEADOWS (8 MILES) On ROUTE 9K
ASH GROVE FARMS
INNISCARA
SARATOGA GOLF CLUB
SARATOGA HOSPITAL
ST. FAITH'S SCHOOL
RECREATION FIELD
PETRIFIED GARDEN
ROUTE 29
BROADALBIN and WEST
AIRPORT
SCHENECTADY BALLSTON SPA
ROUTE 50
HATHORN NO. 3 SPRING
HAYES SPRING
GEYSER PARK
PICNIC GROUNDS
SARATOGA SPA
SPA GOLF COURSE
TREE NURSERY
ALBANY and SOUTH
ROUTE 9
AVENUE OF PINES
BRIDLE PATH
BATH HOUSES
SARATOGA RACEWAY
GRAND UNION HOTEL
INFORMATION CENTER
CASINO
CONGRESS PARK
CONVENTION HALL
CHAMBER OF COMMERCE OFFICES
D. & H. R. R. STATION
CITY HALL
STATE ARMORY
HIGH SCHOOL
HIGH ROCK SPRING
RED SPRING
SKIDMORE COLLEGE
SARATOGA RACING ASSOCIATION
YADDO
THE BROOK
PIPING ROCK
RILEY'S
NEWMAN'S
LAKE LONELY
AVENUE OF POPLARS
SARATOGA LAKE
SARATOGA SULPHUR and MUD BATHS
SCHUYLERVILLE
ROUTE 29 EAST
TO OLD SARATOGA MONUMENT (12 MILES)
TO SARATOGA BATTLEFIELD 10 MILES
ROUTE 9P
TO N. Y. STATE VETERAN'S CAMP GLENS FALLS & NORTH
McGREGOR LINKS
FEB 1952
ARROWHEAD
DIVISION OF MAPS
POINTS OF INTEREST
1. Chamber of Commerce Public Information Center.
2. Congress Park and Memorial Bandstand.
3. The Casino, of Canfield fame, with memories of games of chance. Now a spacious ballroom, museum, and convention center.
4. Grand Union Hotel, of world fame.
5. Skidmore College, a leading university for women.
6. Saratoga Racing Association track, oldest and most beautiful in U. S.
7. Saratoga Raceway, half-mile harness racing oval. Parimutuel night racing.
8. Yaddo, former Spencer Trask estate, now artists' and writers' retreat.
9. Saratoga Lake. Bathing, fishing, boating. Picnic grounds, amusement park, trailer camp. Seaplane landing facilities.
10. Saratoga Sulphur and Mud Baths. Privately owned and operated.
11. Recreation Field. Wading pool, baseball, football, cinder track, skating.
12. High Rock Spring, known to early American Indians as "Medicine Spring."
13. Chamber of Commerce, Saratoga National Bank, 2nd Floor.
14. McGregor Golf Links. 4 miles north of Saratoga. 18 distinctive golf holes you will remember. Private. Guest play welcome.
15. Inniscara, home of the late Chauncey Olcott, celebrated Irish tenor. Private.
16. Saratoga Golf Club, 9 holes, third oldest in United States. Private. Guest play by invitation only.
17. Ash Grove Farms, famous for blooded stock.
18. U. S. Government Fur Animal Experimental Station.
19. Petrified Sea Gardens, one of America's great natural wonders. Privately owned and operated.
20. Saratoga Spa—America's most famous Spa—owned and operated by the State of New York.
A. Hall of Springs. Concerts.
B. Administration Building. Summer Theater. Laboratories.
C. Roosevelt Baths and U. S. Veterans' Facility.
D. Recreation Center, Golf House, Swimming Pool.
E. Gideon Putnam Hotel.
F. Spa Golf Course.
G. Geyser Park, spouters, picnic grounds.
H. Washington and Lincoln mineral water baths.
21. New Community Library.
22. World's largest forest tree nursery. State-owned.
23. Convention Hall. Nominating place of New York Governors. 5,000 capacity.
24. Airport, class 3. 4,000-foot runways.
25. Hathorn No. 1 Spring. Free parking area.
26. Broadway Drink Hall. All State bottled waters. Music.
27. Hathorn No. 3 Spring.
28. St. Faith's School for Girls.
29. Dining and dancing spots for evening entertainment too numerous for listing. Ask at Information Booth.
30. U. S. Post Office.
31. Saratoga Alpine Ski Meadows. 6 Ski Tows. Platter Pull ski lift.
32. Community Theatre.
33. Congress Theatre.
● Indicates Traffic Light.
FOR INFORMATION ON HOTELS, GUEST HOUSES AND RESTAURANTS:
Write for Chamber of Commerce List or Ask at Information Center
BUS SERVICE PROVIDES EASY ACCESS TO ALL PARTS OF THE CITY

DAVID BAKALAR

Drawing from his extensive background in metallurgy and technology, David Bakalar creates sculptures that address the relationship between science and ourselves. With genetic and DNA research now at the forefront of public discussion, Bakalar uses the outline of man contained within a double helix to represent our contemporary state of being. The quest to completely map the human genome is treated like a spin of the roulette wheel: exciting and potentially dangerous, with the outcome unknown.

Wheel of Fortune
1988

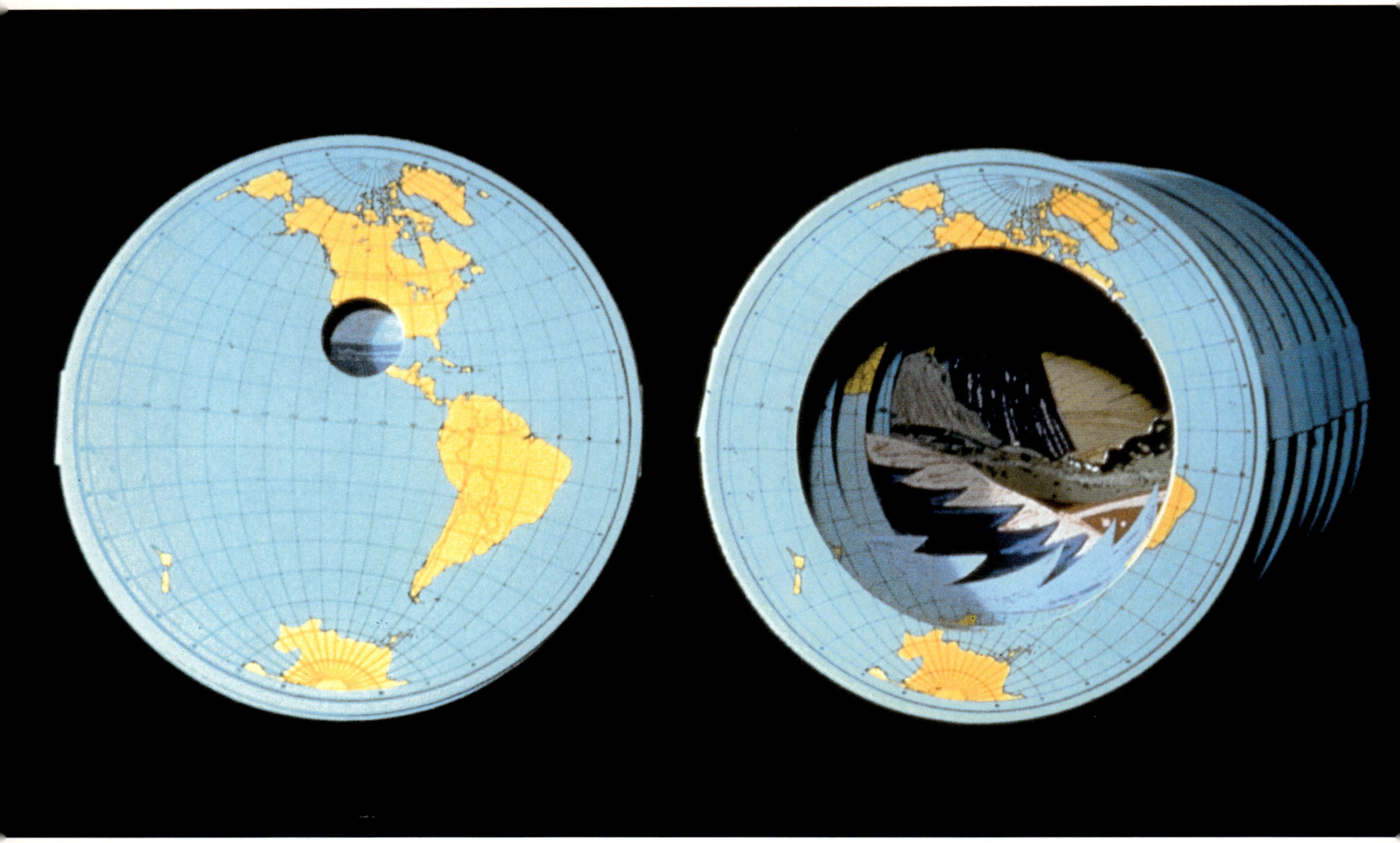

ABOVE
Tunnel Map
1988

CAROL BARTON

Carol Barton has used the symbols and imagery of map making to explore themes of travel, space and time in her editions. The time-based nature of a book – information is revealed as one moves through the pages – allows Barton to take viewers on a journey. These books take the form of a tunnel book, which offers information layer by layer. The layers are revealed as one opens the intricately constructed object and views it from several vantagepoints. In these books the maps and fragments of maps take on a sculptural presence as components pop out and physically extend into space.

THÉODORE COULOMBE + ALAN KRATHAUS

The work of Coulombe and Krathaus is concerned with the relationship between the claim of conquest and actual control. Past projects by the two have examined Polar exploration and the work of early Geological Surveys in the American West. This project co-opts the image of the U.S. Army Signal Corps, charged with all aspects of documentation, communication, and propaganda. The artists place the ostensible truth of Signal Corps documentation hard against a reality that contradicts this implicit claim of military control.

LEFT
Air to Ground Triangulation. Bridgehampton, N.Y. E.T.O. U.S.A. 12.13.94
From Signal Corps Series
1994

OPPOSITE
Beach Head. Bridgehampton, N.Y. E.T.O. U.S.A 12.13.94
From Signal Corps Series
1994

SIGNAL CORPS U.S.A.
BEACH HEAD. BRIDGEHAMPTON, N.Y. E.T.O. U.S.A. 12.13.94

WIM DELVOYE

An artist who works in diverse media and tackles a wide range of cultural issues in daring and controversial ways, Wim Delvoye attempts to reveal cultural peculiarities in his art. Within his extensive body of work, Delvoye has produced a series of forty imaginary maps that use clever wordplay, sly references to human anatomy and recognizable forms for landmasses and bodies of water.

RIGHT
South Vikxonia
1992

IZUCH
POLOPORS
QOROS
HILVEP
REGOR
ULEVEL
AJILDO
DOUZT
PODON
USP
KOLBOD
KOLBOD BAY
PORON
BLOGO
MIATOR
MOOXY
NEBUK
ORONO
EMALE
TEODI
STUSEF
TRICE
OLZI
OJI
INAEP
PANCISES
SISEN-AAN
ETIFN
STESTAF
MOAIS
TANTEPHAO
LIRVIC
NEW-SOA
LIRUTICO OCEAN
SIAXY
PHAGOA
XUXTO
OTIE BAY
PENIAO
FITACIO
ILE SOABEN
ILEASON
UKOR
AOVEG
CHORRI
AZAYS
UMBROT
IRT
LEEO
SEA
OSXUSU
SONCERI
JAAIK
KBOL
WATPUPU
ERTENX
ITEKALR
OOST-KU
YOIX
PJIA
SAEBX
SENP
OMAZIO
ACOS
TYGRIS
BOBEL
OONL
GGYRIC
KVOIP
BESTX
VOTBAGO
SIONSP
UXSPON
SIA SOAI
JARATA
AKARTO
BE-IOR-IOS
ANTOHO
PEEISS
UTTISCHE ZEE
SEE OF FIOX
BELHOMO
KOSTO
VUNOK
OTTUN
OTONO
NORA
EAST-ETCI
ILESPES
DECIJTN
FIOX
CRISXI
VINSTRAE
OHORGOS
TOROS
DOMORI
THBESTH
KUWERVI
IROUP
CRALIN
WIRON
SUYT-PE
TANPH
AINTX
AMMELONCK
NISACK
SWANTIT
KIAIO
SRODI
ETSOAL
FAESIS
ASIJAGR
JURAX
LAKARTSO
TONX
ELVANIË
XIPSE
BRANERIS
EVOLEN
ANTAPANE
ASISO KANAAL
SCATALO
ODORIWR
XRUBIJE
OOST-TINSEL
SILEAX
GIOVEORD
WIDADIA
ASISO
BOLIX
SRODI ISLANDS
OJI BAAI
BRENICEX
ITOSEN

EXIT

SAM EASTERSON

Sam Easterson recently created the *Animal, Vegetable, and Video Project*: an undertaking devoted to compiling the world's largest library of video footage recorded from the perspectives of animals and plants. In this installation, Easterson outfitted six indigenous species to the island of Manhattan to imagine what a map might have looked like a thousand years ago. Stuffed models of the animals – such as a deer and mouse – are placed within a topographical layout of the island made from extension cords that connect the components of the piece. This presents the Manhattan of a distant past, populated by animals instead of people, when trees in lieu of skyscrapers dominated the skyline.

OPPOSITE & RIGHT
Animal, Vegetable, Video: Memories of Manhattan from a Millennium Ago
2000

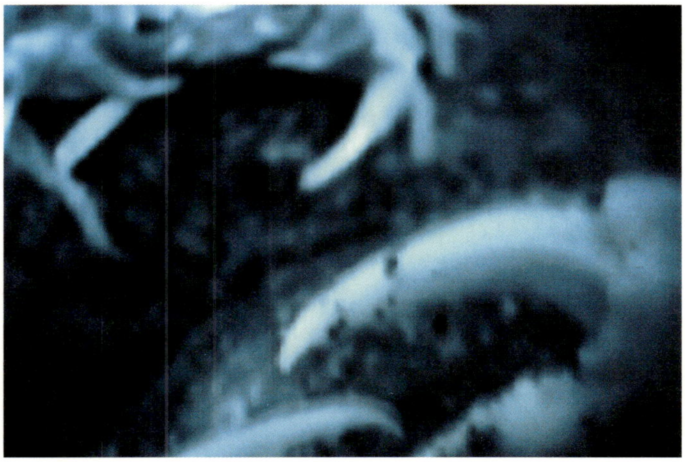

KATE ERICSON + MEL ZIEGLER

Ericson and Ziegler often focused their work on the political, economic and historical circumstances that quietly surrounds us. These 173 jars, filled with Benjamin Moore latex paint, match the colors found on the exterior walls of the buildings and monuments surrounding the mall in Washington, D.C. The jars are installed on the wall according to their location, and each is sandblasted with the commercial name given to the color. In the nation's seat of power, the colorful names given to shades of paint take on poetic and ominous layers of meaning.

LEFT
Dark on That Whiteness
1987

OPPOSITE
***Dark on That Whiteness** (detail)*
1987

NATURELLE
WALNUT
CREAM
MORNING
MIST
WALNUT
CREAM
DAWN
GLOW

NINA KATCHADOURIAN

Nina Katchadourian's cutout maps are made by painstakingly separating roads from their backgrounds, so that only the pathways themselves remain. In her *Handheld Subway*, for example, she removes the color coded lines from a New York City subway map, and photographs the kaleidoscopic jumble in the palm of her hand. This changes the monumental subway system into a delicate and personal web of paper. The arterial-like form left after removing the United States highway system from the rest of the country further illustrates Katchadourian's unique interpretation of an ordered network. We see the ephemeral nature of maps, the fragility of information and our surroundings, and how information's properties can change in different contexts.

RIGHT
Finland's Longest Road
1999

OPPOSITE
Austria
1997

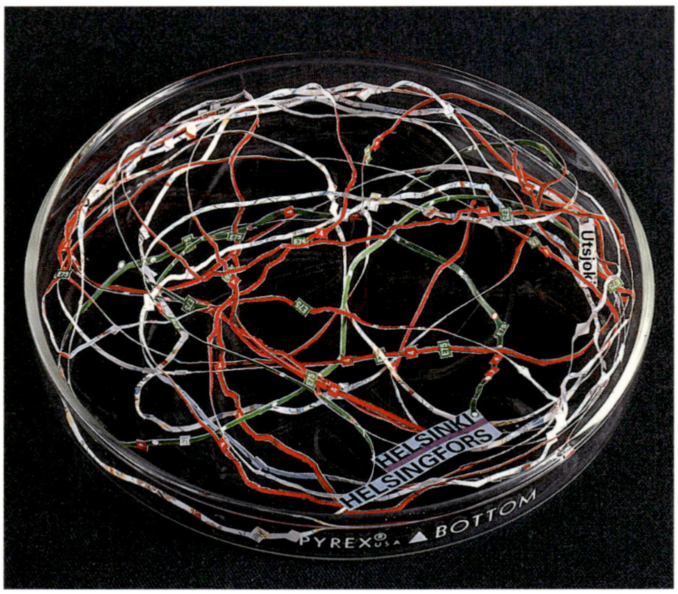

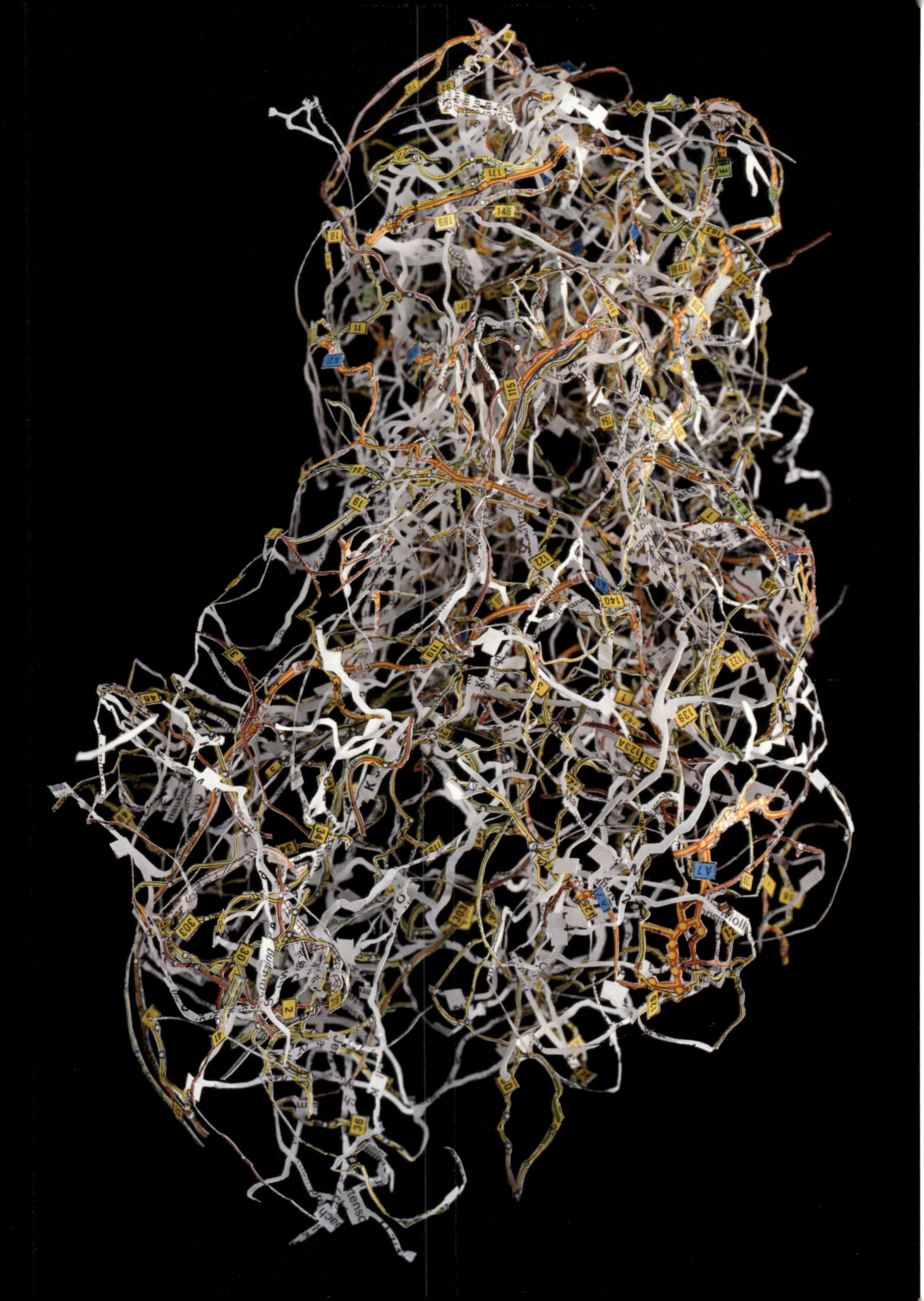

JOYCE KOZLOFF

In striking color and fine detail, Joyce Kozloff has transcribed old maps and globes, highlighting the cultural circumstances that maps and changing boundaries often reveal. She exposes the allure of power by representing conquest. In her newest work, *Targets*, Kozloff has constructed a walk-in globe which surrounds viewers with painted sections taken from United States military maps of countries the U.S. has targeted since World War II. Inside the globe, circumstances are reversed, and we become a small figure amidst looming enemies.

LEFT
Knowledge #75: 1290
1999

OPPOSITE
Targets
2000

CONGO
KINSHASA
ZAIRE
TRIPOLI
LIBYA
SUDAN
CAMBODIA
KUWAIT
IRAQ

SELF-PORT
DIVIDED PROPOR
BETWEEN THIS BLAC
REPRESENTING LIFE
AND THE REMAININ
SPACE REPR
COME, BASE
STICAL LIFE

MICAH LEXIER

Micah Lexier's self portrait is based on a statistical equation he created in 1998 to decribe life he had lived versus his life yet to come. Lexier fills the wall with bold vinyl text and as the letters represent his life at that moment in time, we see ourselves in the arrangement of words and imagine what we would look like as this wall. How do we represent the measure of our time and experience? This poetic, site-specific piece, with its proportions changing at every installation, soberly reminds us of time passing and suggests a new way of representing a life.

OPPOSITE AND ABOVE
Self-portrait as a wall divided proportionally between this black type representing life lived and the remaining white space representing life to come, based on statistical life expectancy
1998

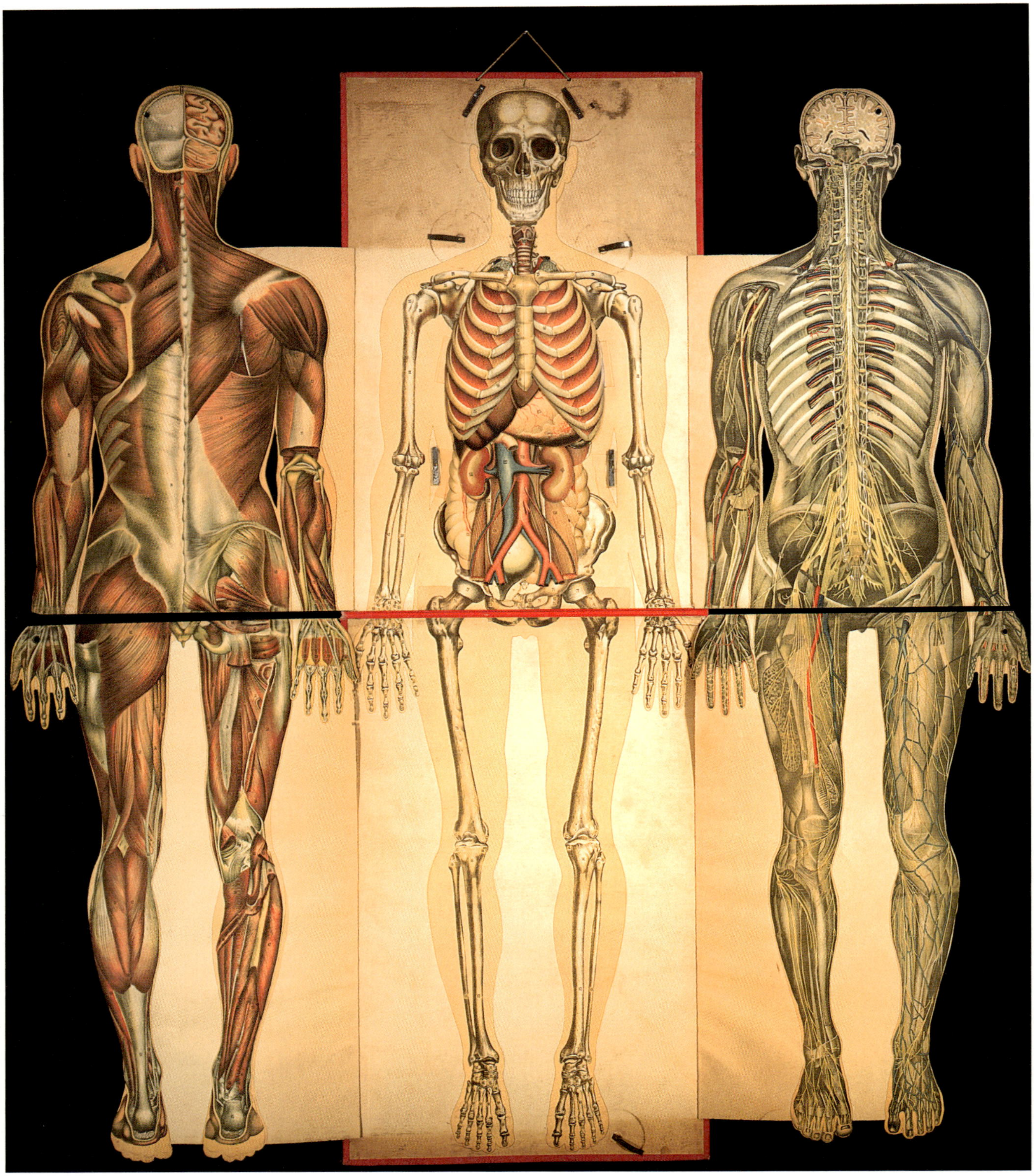

ABOVE

Pilz Anatomical Manikin

c. 1900

Re-Presenting the Body... Anatomical Models as Maps

RICHARD WILKINSON

ANATOMICAL MODELS AND ILLUSTRATIONS SERVE A WIDE VARIETY OF FUNCTIONS AND HAVE many intentions. The obvious intent, although rarely stated as such, is to serve as a map: where to find certain biological features, what other features are spatially related to the one of interest, how these features function relative to each other. As with other maps, illustrations of the human body represent to us those things which are otherwise "insensible," either because they cannot be seen, we cannot get access to the source, or cannot be understood without symbolic representation. The arteries of the lungs, for example, cannot be seen, we cannot obtain lungs to try to see them, and they can only be visualized by schematic representations, as the network of vessels is too complex, too small and too inaccessible to be visualized directly.

The history of anatomical illustration indicates the progression of the intended purposes of the illustrations. The earliest representations – at least those we can recognize as such – most likely were direct efforts to objectify the anatomy. Unfortunately, these early efforts such as those of Galen in the second century were hampered by the lack of human subjects. Galen's anatomy, then, was a blend of human and non-human sources. As in many other areas of science and mathematics, anatomical illustration and knowledge in the Western world was kept alive by Arabic speakers, to be 'rediscovered' during the European Renaissance.

Leonardo da Vinci no doubt represents Renaissance anatomy to most of us, but the premier anatomist of the period was *VESALIUS*, whose major works (***De humani corporis fabrica*** and ***Epitome***) appeared in the mid-sixteenth century. Joining other Renaissance luminaries in Padua,

RIGHT
ANDREAS VESALIUS
Skeleton, Posterior View from De Humani Corporis Fabrica, Book 1,
1543 (printed 1934)

BELOW
JOHANNES DE KETHAM
Fascilius medicinae in quo contintentur: videlict
c. 1495

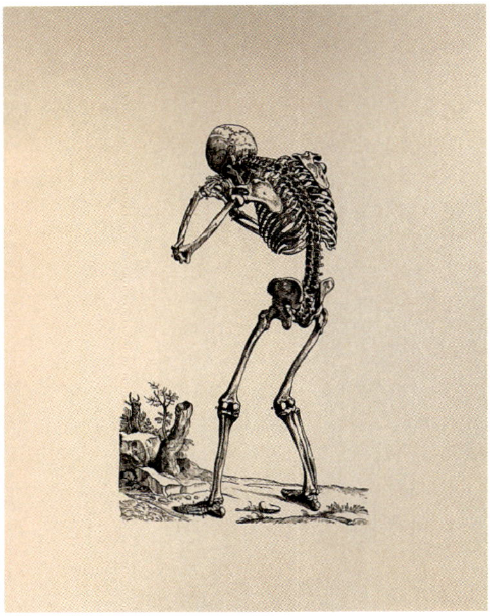

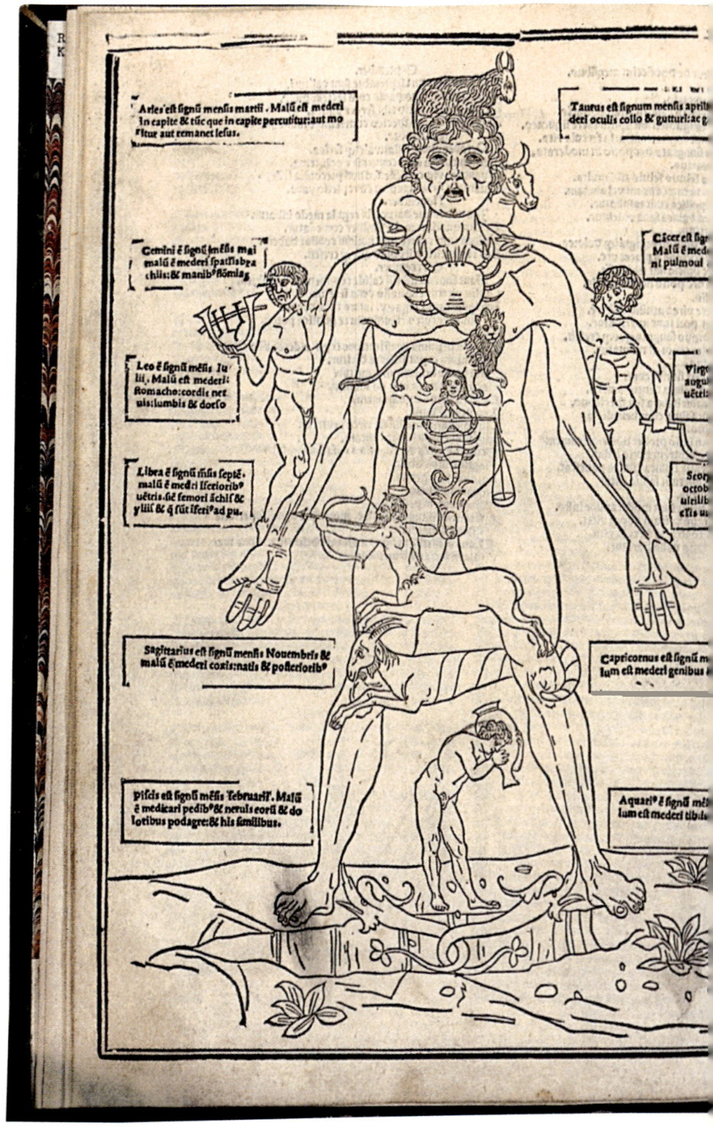

VESALIUS sought to correct the errors of galenic anatomy. Unlike Galen, *VESALIUS* utilized human cadavers, allowing a vast improvement in understanding human anatomy [*VESALIUS* boasted of correcting over 200 of Galen's errors]. *VESALIUS* was a dissectionist; he sought to understand the reality of human anatomy by uncovering it, layer by layer. The piece on display in the Mapping exhibit is one of the classic *VESALIUS* poses, showing the human skeleton from the rear. The skeleton is posed in a clearly life-like, emotional position, illustrating not only the anatomy via a bone map, but also representing the map as a living person, and not a cadaver. *VESALIUS* was not an artist, however, and his illustrations were probably done by at least two artists. The most likely attribution of the rear skeletal view is to Venetian, a student of Titian. *VESALIUS* himself utilized and drew sketches of the anatomy in his pedagogical efforts, but none has survived.

Skeletal drawings such as this one represent the end of a spectrum or series. Other drawings in the series show increasingly more soft tissue — muscles, organs — and the illustrations are presented in stages, according to the dissection series. The illustration immediately preceding the skeletal drawing, for example, would show some of the deeper muscles.

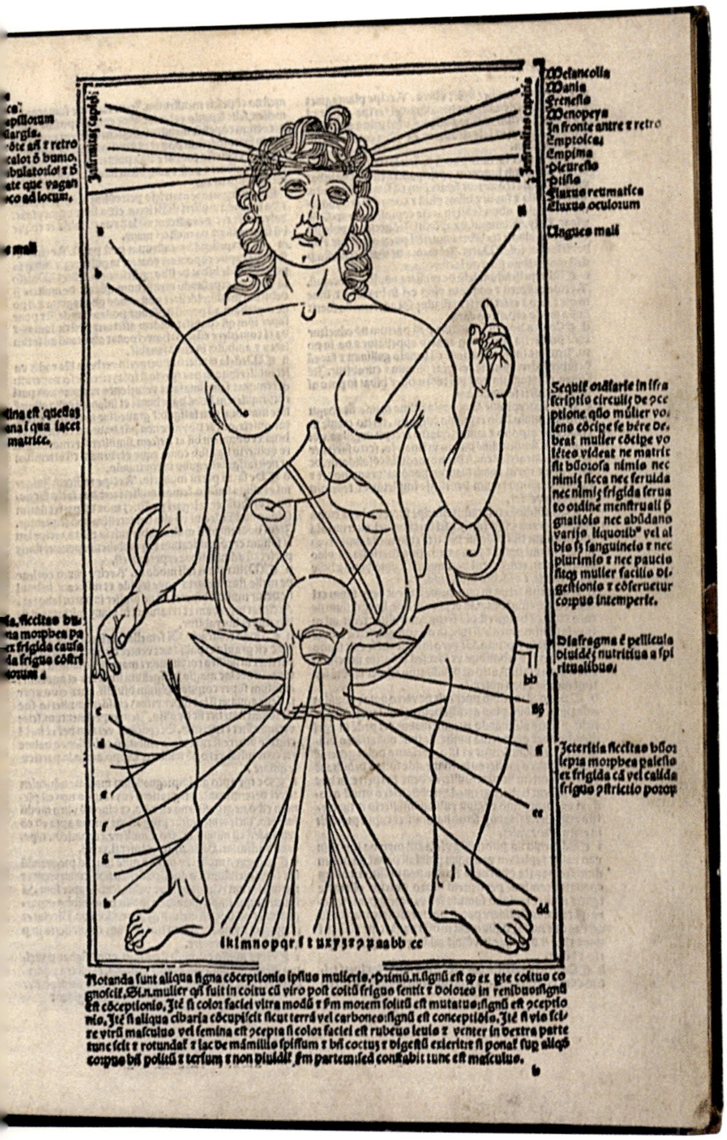

This layer-by-layer approach finds clear application in subsequent centuries of illustrations, reappearing as "flap art" as seen in the ***Pilz Anatomical Manikin*** (pg. 46) and on the ***Four Seasons CD ROM*** produced by the Duke University School of Medicine. These methods of illustration represent simple technological solutions to complex three-dimensional problems, as the observer peels away layers to reveal deeper maps. Other, contemporary approaches to this same genre of illustrations are the layered transparency maps, in which anatomical layers are added or subtracted by the use of large transparent flaps.

The ***Pilz manikin*** (c. 1900) is one of the few remaining examples of this genre of utilized anatomical maps. Hung in physicians' offices or carried with the doctor on house calls and/or hospital visits, these diagrams allowed the representation of various levels of anatomy as the physician explained the subject matter to the patients or students. The utility of these manikins is responsible for their relative rarity, as they were subject to destruction and loss of parts. Hundreds of years earlier, "flap anatomy" appears in forms such as that represented by the early 17th century "***Four Seasons***" illustrations. The anatomy depicted is essentially Vesalian, and the flaps allow the observer to discover different layers and different anatomical systems. The ***Four Seasons*** piece is, of course, much more than anatomy, as the text and illustrations deal with geography, trade, astronomy, urology and horticulture, to name a few. The illustration is attributed to Robert Fludd (1574-1637), and is seen as primarily a device to assist in the practice of urology (Cazort et al. 1996). Taken in its entirety, however, it is much more than that, and provides a window on science and art in the early seventeenth century in Europe.

Carrying the flap anatomy concept even farther, modern computer technology is brought to bear on the Visible Human Project, as illustrated in the CD ROM ***About Faces***, provided by *HT MEDICAL SYSTEMS*. Here thin flaps are electronically peeled back as the observer manipulates the image with a computer 'mouse.' Each movement of the controller pushes us deeper into the cranial or facial anatomy, as if thin flaps had been removed from the body. Despite the relatively elaborate technology, we are still presented with a two dimensional view of a three dimensional object, and we are still dealing with representation, not reality; we have a map, not a body [or landmass, to go to the geographic use of a map]. Unlike a "real" map, the computerized dissection has no roads, no elevations, no key, and it is corresponding difficult to find one's way through the landscape of the interior of the head.

In addition to helping us find our way through the human body, anatomical illustrations also provide ancillary information, revealing the "mapmakers" intention, culturally-specific concerns and a variety of biases. Despite attempts at accuracy and/or emphasis, these various illustrations are often as subjective or topic-specific as cartographic illustrations. Two early pieces in the Mapping exhibit demonstrate this point quite well. In *KETHAM*'s late 15th century illustrations of a man and a woman (pg. 48-49), the anatomy is overshadowed by the obvious concern with astronomy in the male illustration, and reproduction for the female. The post-Enlightenment illustrations of *ALBINUS* (1697-1770) eliminate references to astronomy, Adam and Eve and most other seemingly non-essential referents, but the artist Jan Wandelaar added life-like backgrounds to accentuate the concept that the illustrations were not of dead people.

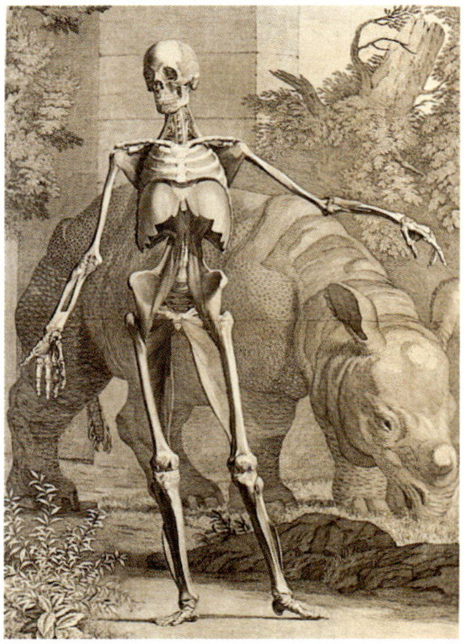

The *ALBINUS* figures are extremely accurate maps of the skeletal and muscular systems, superimposed on an interesting background which demands some contemplation. The background may be seen as equivalent to the marginal illustrations in many of the early maps in the exhibit, illustrations which emphasize the purpose of the map or make general comments about the concerns of the mapmapker's culture. In the two *ALBINUS* figures presented in our exhibit, a rhinoceros looms inexplicably in the background. Five years prior to the publication of ***Tabulae sceleti*** in 1747, the first rhinoceros was introduced to Europe from India. The rhinoceros was understandably of interest to Europeans, and its inclusion is most likely reflective of Wandelaar's wish to 'enliven' the illustrations with contemporary referents and living objects. The foregrounded skeleton is an excellent map of that anatom-

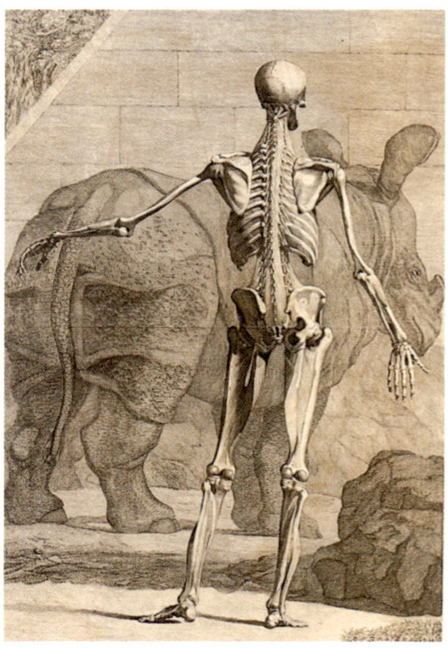

RIGHT
JEAN-BAPTISTE MARC BOURGERY
Volume 6 of Traité complet de l'anatomie de l'homme
1852

OPPOSITE TOP & BOTTOM
BERNHARD SIEGRFIED ALBINUS
Tabulae sceleti et musculorum corporis humani
1747

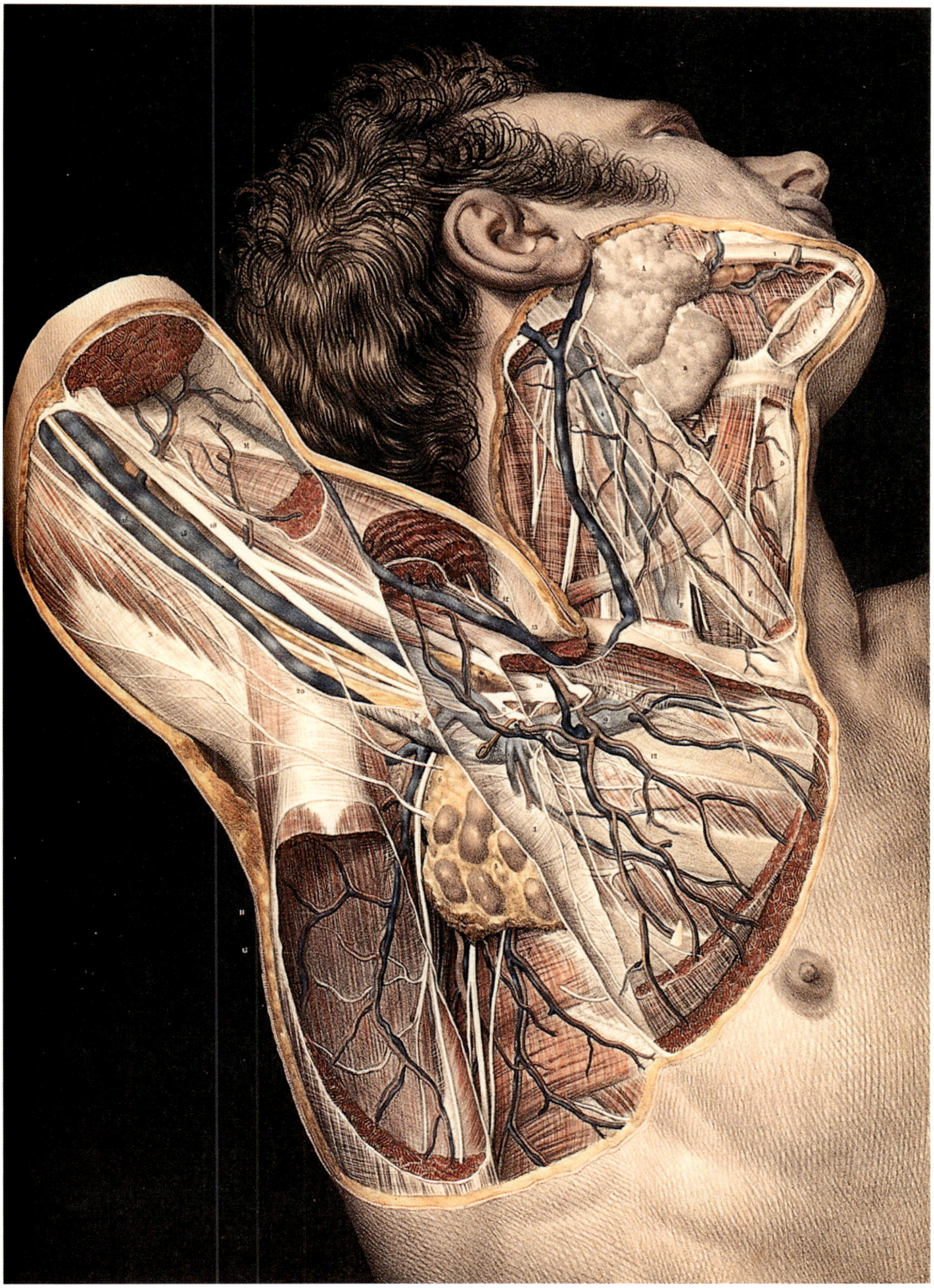

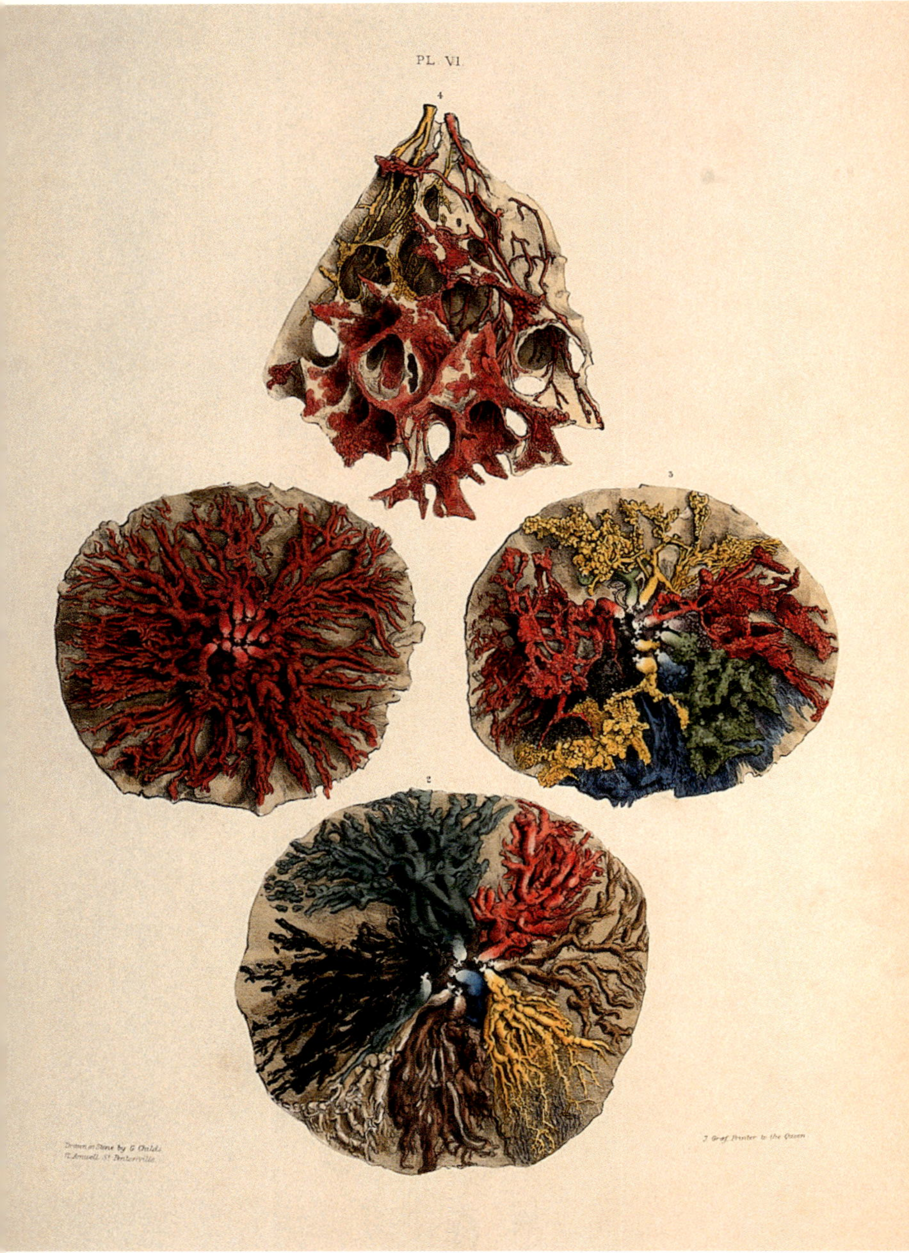

LEFT
ASTLEY PASTON COOPER
On The Anatomy of the Breast
1840

ical system, and *ALBINUS* went to great lengths to insure its accuracy. *ALBINUS* created a two-phase grid system of ropes and string, allowing the artist a constant reference at varying distances from the subject. Applying a grid system to the drawing project not only improved the accuracy, but for our purposes it emphasizes the mapping element of the project. The subject itself was a product of intensive effort at accuracy, as *ALBINUS* procured a "perfect" 25 year-old cadaver, and went to great lengths to position the resulting skeleton in life-like poses.

One of the more obvious map-like anatomical illustrations in the exhibit is a drawing of the arterial system in one half of the human body, done by *W.H. THOMPSON* in 1858 (back cover). Arranged as it is in a linear, "north-south" alignment, the drawing bears obvious resemblance to a riverine system such as the Hudson River, with smaller arteries representing the many tributaries of the main system. In addition to the geographical similarities of this anatomical drawing, it also has the maplike quality of simplification of a complex system, and the concentration on a specific aspect of a much broader landscape.

Another anatomical map which concentrates our attention on a single element is *ASTLEY PASTON COOPER*'s ***Milk Ducts in the Breast***. *COOPER*

(1768–1841) was a royal physician, anatomical author, observer, dissector and collector. To illustrate the structure of the milk ducts in breasts *COOPER* injected the ducts with colored wax, then had his illustrators produce lithographs of the observed structures, and then the lithographs were hand colored, resulting in the vibrantly colorful and lava-like representation of this biological system. The mid-19th century was a time of extensive use of hand-colored lithography in anatomical illustrations, and one of the more spectacular versions of this technique is found in the 8 volume ***Traité complet de l'anatomie de l'homme***, by *JEAN-BAPTISTE MARC BOURGERY* (pg. 51).

The illustration of the axilla and neck was done by *NICOLAS-HENRI JACOB*, and appears in volume 6 of *BOURGERY*'s massive treatise, published in 1852. This lithograph is a dramatic "clean" drawing of the anatomy of the neck and axilla, and contrasts markedly with photography of the same area [of cadavers]. The illustration is described as among the most lavishly illustrated anatomical and surgical treatises ever published and is a beautiful example of the then-new art of colored lithography. In attempting to reveal the "true" anatomy, the artist goes far beyond that possible with photographs of the 'real thing' a century later. The use of colors for the different parts of the body – vessels, nerves, bone, muscle – provides more visual relief to this complex landscape, and therefore allows us to model that which cannot be otherwise sensed with such clarity. The artist also adopted a technique having ancient roots: illustrating the "model" as alive and well, thereby emphasizing life over death and offering another contrast with contemporary photography of cadavers.

In addition to the two dimensional anatomical illustrations discussed above, the exhibit features several three dimensional objects which serve to locate and relate various aspects of our anatomical beings. The mid-nineteenth century phrenological head is a globe-like map of the various places on the surface of the skull where specific behaviors, tendencies and characteristics were thought to reside. A phrenologist, or "examiner" as they preferred to be called, would evaluate a person by examining specific locations on the skull, searching for different levels of development. In order to maintain the impression of science, the examiner needed to score exactly the right place on the skull for each characteristic or behavior, and this was made possible by the precise maps which were engraved on plaster – and sometimes real – skulls. In addition to the models, phrenologists had recourse to books which guided their examinations, and provided a discourse on the findings. Our exhibit includes such a book, ***Wells' Descriptive Chart***, which was used by the phrenologist *F.E. ASPINWALL* to examine his wife Rebecca in 1875. Rebecca came out fairly well on this examination, although she was admonished to work on those weaker parts of herself which were due to her gender.

Joining the phrenology materials in the display case are a real globe and a small human head with acupuncture signs and channels. The globe was in the Ft. Edward school system before being accessioned by the New York State Museum, and in addition to drawing attention to the globe-like qualities of the two adjacent simulated skulls, the globe also captures the theme of the exhibit in boasting of its "improved" qualities, which went to the extreme of announcing "new constellations." This new and improved map could take one through the heavens and, considering the astrological implications, perhaps beyond.

A small vinyl head with numerous acupuncture points and a series of lines (meridians) is a useful and elegant map to guide the proper placement of needles. As described by one author, "Needles are placed at specific points along meridians, or channels. These channels are like rivers with tributaries that flow into increasingly narrow rivulets, mimicking nature's flow of water to increasingly smaller streams. The twelve main meridians, like the twelve main rivers of China, represent an internal system of communication and transport, just as actual waterways permit communication and transport in the outer world. The human body is viewed as a miniature model, or microcosm, of the universe." (Cassileth 1998)

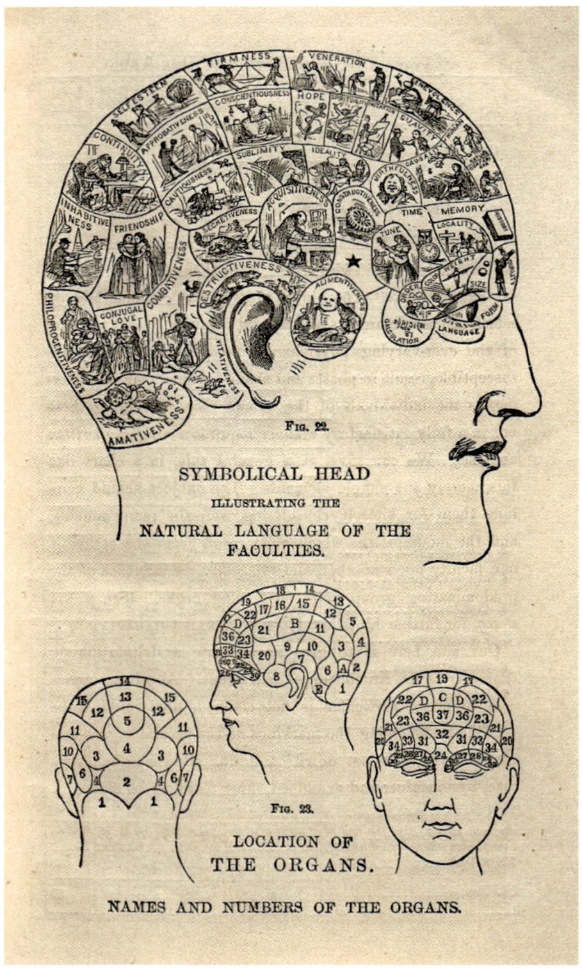

Also in the three-dimensional genre are the two carved mannequins, representing a man and a pregnant woman. These small figures have removable abdomens, revealing simplified aspects of internal gastrointestinal and, in the case of the woman, reproductive anatomy. The woman comes with a fetus, attached to a placenta with a silk thread, all of which is beneath the intestines as one lifts successive layers of ivory anatomy. The figures reveal, then, relative locations and appearances of some otherwise invisible anatomical features. The actual purpose of the figurines is unknown. At this time, and earlier, similar figurines of women were utilized by physicians in China to diagnose illness in their female patients, as they were not allowed to examine the women themselves. These figurines served as surrogates, on which the women's servants could locate the areas of concern for the physician. Since this form of culturally-mandated modesty was not part of the European cultural tradition, these European mannequins must have served other purposes, as yet unknown. Given the obvious reproductive emphasis, coupled with the anatomical simplicity (making it unlikely they were used for any kind of medical education), the figurines may have been nuptial gifts given to the relatively well-to-do (and sexually inexperienced) bride and groom. ⊕

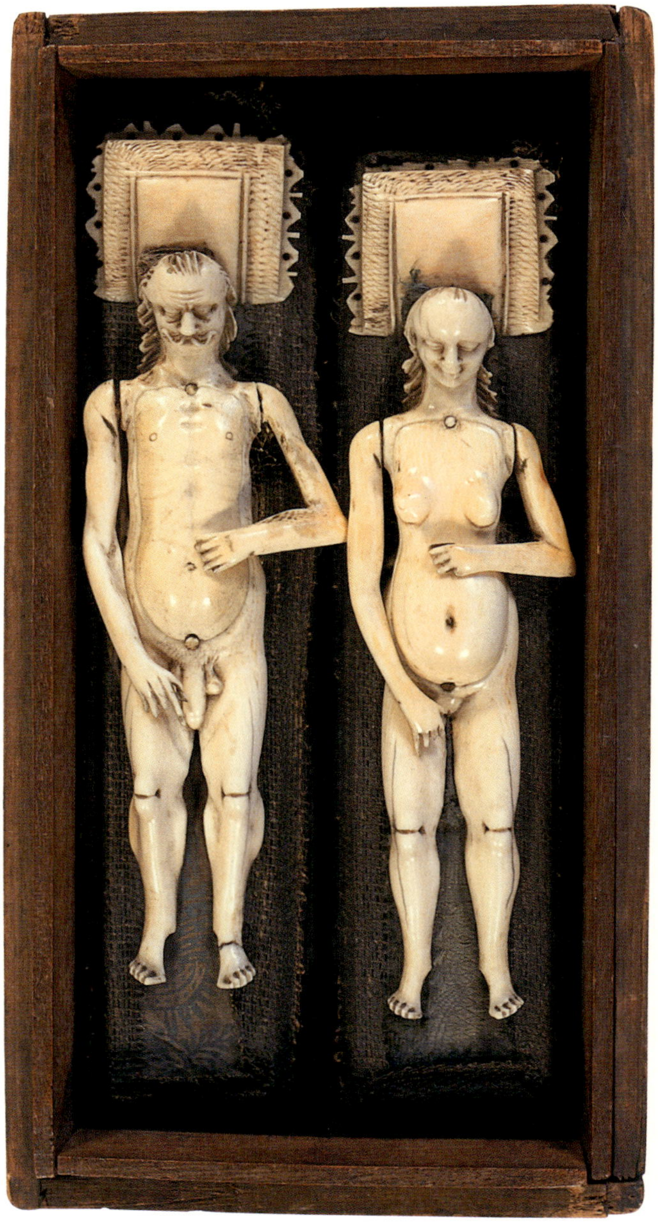

RIGHT
European Medical Mannequins
18th century

OPPOSITE LEFT
Phrenology Head
c. 1850–1900

OPPOSITE RIGHT
Wells' New Descriptive Chart
1787

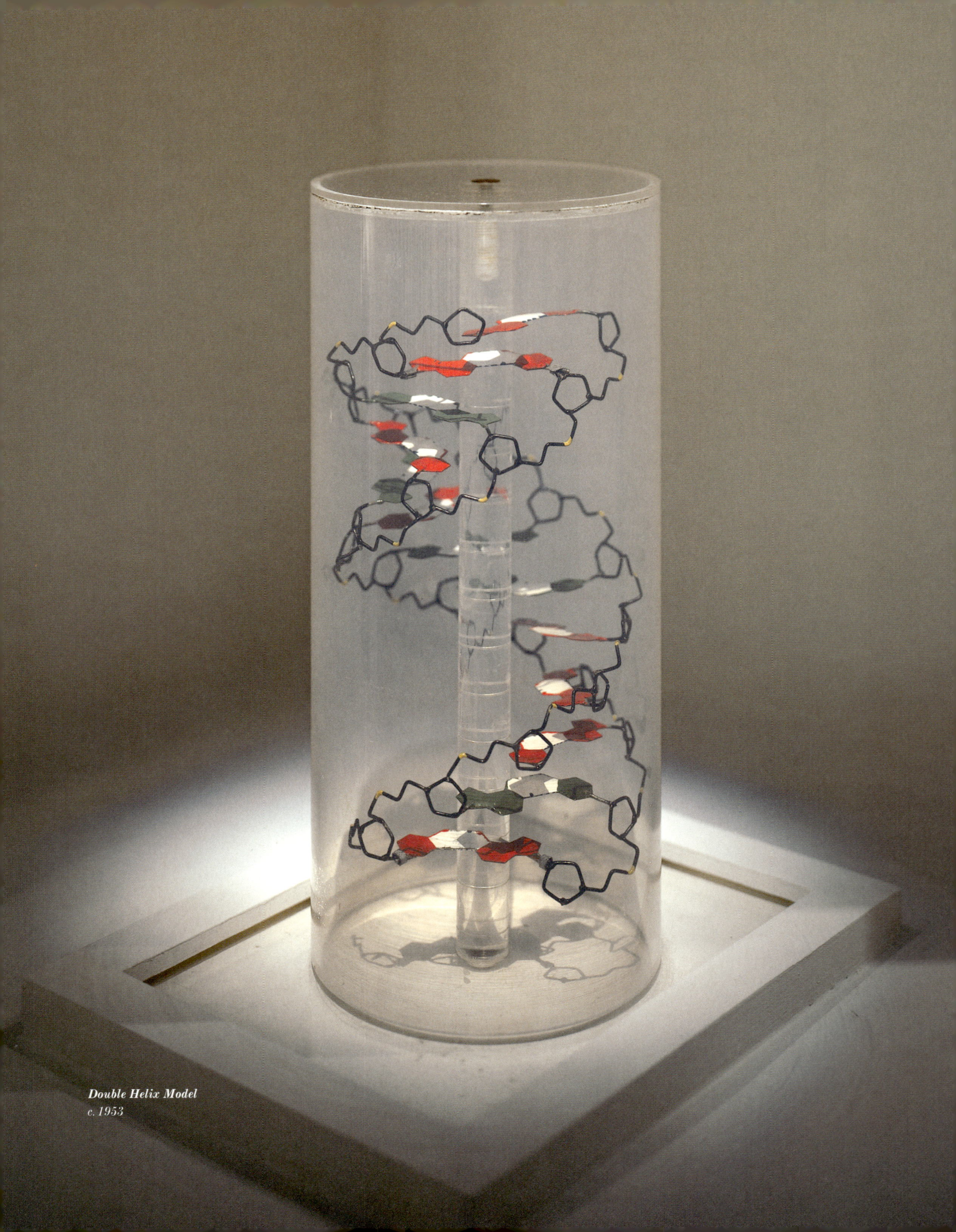

Double Helix Model
c. 1953

Genetic Mapping: I Map, therefore I Am

BERNARD POSSIDENTE

A MAP, IN A SIMPLE WAY, IS A SYMBOLIC DEVICE THAT WE CAN INTERPRET TO FIND OUR WAY FROM one place to another. Organisms must have a map that tells them how to find their way from one generation to the next. Very little of the human organism, for example, is present in a fertilized egg, but a complete set of genetic information is contained within it, and this is what insures that a human organism will develop. Such a map is carried within every living cell of every organism. The science of genetics is the study of these maps and how they guide the development of each organism every generation, and preserve the essence of each organism over evolutionary time. For four billion years, since the origin of life ancestral to our own, the information in genetic maps has guided the generation of all kinds of organisms, in biological space, through complex pathways back to their own reproduction. Each new generation ends up in the same place on the genetic map as its ancestors, except for mutations that may alter the map.

Over 2,000 years ago, Aristotle anticipated our current understanding of genetics and development in his ***De Generatione Animalium***. There he proposed that a force contributed by the male parent shapes, animates and 'generates' a living recreation of the parents from a formless substance contributed by the female. Although today we know that both male and female parents contribute matching sets of genetic information, his concept of the organism being generated and shaped by information from the same species of parents is much closer to our current scientific understanding of life than competing ideas that were commonly held before the modern science of genetics emerged.

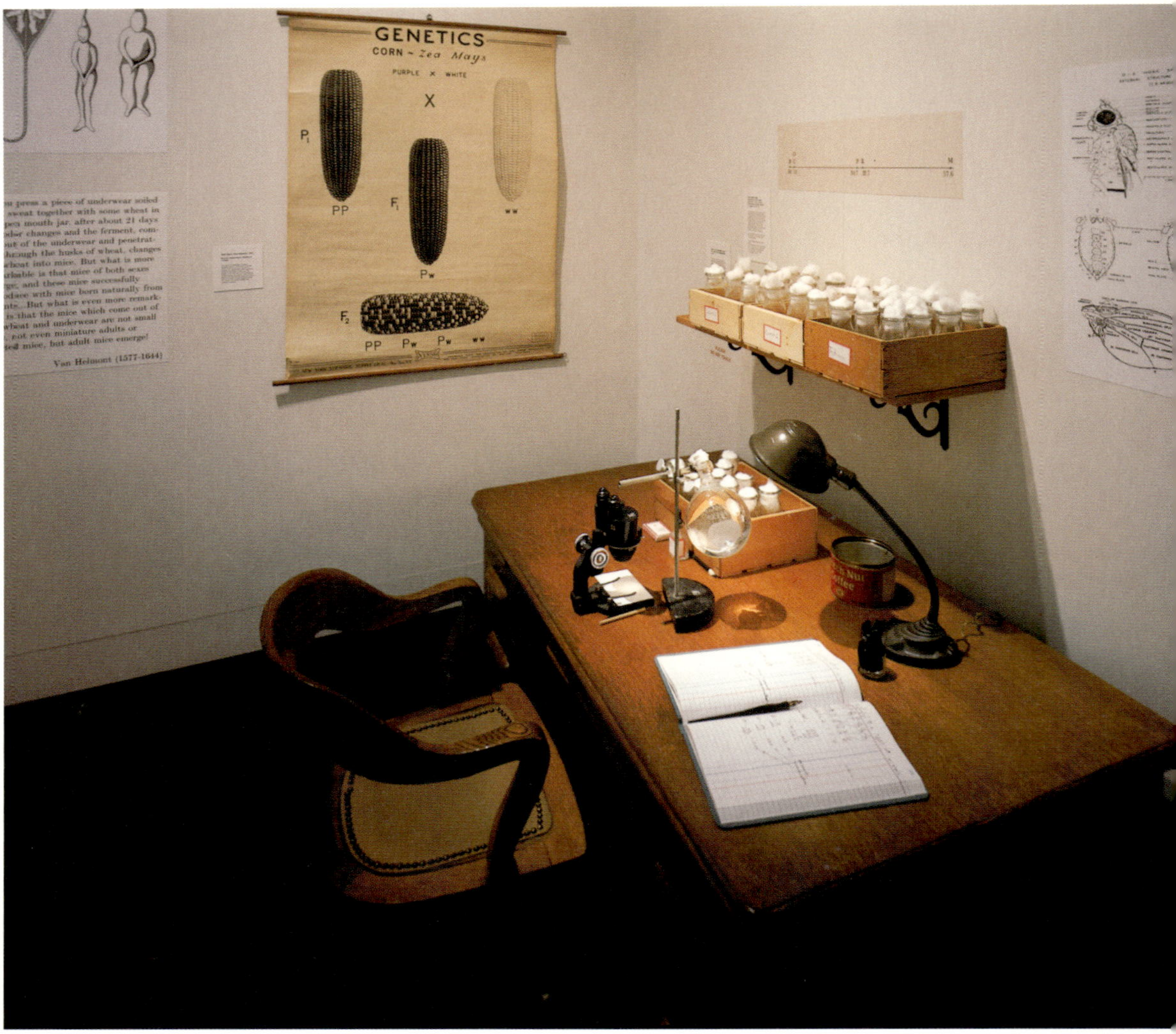

One of the competing ideas, known as 'emboitment,' was that inside each organism there is a tiny organism waiting to pop out, and inside that one is another, and so on. Coexisting with emboitment theories was the ancient idea of 'spontaneous generation,' which had, by the Renaissance, been elaborated into virtual recipes for generating particular kinds of organisms.

Experimental evidence began to accumulate during the Renaissance, particularly after the invention of microscopes, that disproved key elements of the emboitment and spontaneous generation theories. Detailed microscopic observation of the transformation of fertilized eggs through embryonic stages of development, recognition that life is cellular, and the shift from typological to evolutionary perspectives on the origin and history of life, were key concepts that precipitated the birth of biology as a new science. These new perspectives helped focus the study of development and evolution on the question of exactly what organizes the generation of an individual organism from a fertilized egg, and how it is passed on from generation to generation, largely intact, but with the potential for adaptive evolutionary changes.

The modern science of genetics began in 1865, when an Austrian monk, named Gregor Mendel, discovered that genetic information exists in the form of hereditary elements that we now call 'genes.' Although Mendel's theory of heredity was published in 1865, it was largely ignored until it was rediscovered in the year 1900. In 1911, Alfred H. Sturtevant, an undergraduate at Columbia University working in the lab of the developmental geneticist Thomas Hunt Morgan, constructed our first genetic map, showing the location of six genes on the 'X' chromosome of a fruitfly.

Once genes could be mapped to discrete physical locations on chromosomes, attention began to turn to understanding the molecular structure of genetic information and exactly how it functions in a cell. Investigations into the molecular structure of genetic information focused on two candidate types of molecules: proteins vs. nucleic acids. *JAMES WATSON* and *FRANCIS CRICK* were among the first molecular biologists who were convinced that genes were composed of nucleic acid, and they caused a sensation in 1953 when they proposed a molecular structure for deoxyribonucleic acid (DNA) as the material basis for genetic information (pg. 56).

OPPOSITE PAGE
Early 20th Century Fruitfly Genetics Laboratory
c. 1913

Including the following reproductions:

Homunculus
c. 1596
after Delenpatius

Recipe for Mice
c. 1620,
after Van Helmont

CALVIN BRIDGES
Fruitfly Anatomy
c. 1910, Drosophila Information Service Study Guide

Research Notebook Page of Genetic Mapping Data
c. 1911–1913

First Genetic Map,
from original publication in Journal of Experimental Zoology, Volume 14
pages 43–59, 1913

Journal of Experimental Zoology, Volume 14
1913

Wall Chart: Corn Genetics
1941

By 1965, the genetic code used by cellular life was decoded for the first time, by something other than a ribosome, to reveal how the information stored in DNA is mapped into the molecular composition of proteins.

The four molecular components of DNA that carry genetic information can be thought of as a genetic alphabet of four letters (A, T, G and C), which are arranged in linear sequences along each chromosome. When genetic information is required, a copy of just the genetic information needed is generated in the cell, in the form of a similar molecule called RNA. Some forms of RNA are, themselves, the functional gene product, but other forms contain a genetic code used by the cell to synthesize sequences of amino acids to form proteins. The RNA letters are read, in the cell, in triplets corresponding to 'words,' which code for the amino acid components of proteins. The sequence of letters in the DNA 'spells out' the sequence of amino acids for a specific protein, much like a recipe. The proteins, once made, generate cellular structures and metabolic pathways, transport materials, communicate information etc., and ultimately, reproduce another cell complete with an entire copy of the same genetic information. This simple system generates, essentially, all the proteins in all the cells of all organisms examined so far. This is why genetic engineering manipulations, developed only a decade later in the early 1970's, allow the transfer of genetic information from one organism to another, or from synthetic genetic code into cells, because at the cellular level there is a universal genetic code (with a few dialects here and there) for translating DNA sequences into amino acid sequences of proteins.

Genetic engineering methods have been instrumental as a powerful research tool for genetic analysis, and their use in applied genetic technology has also developed rapidly along with our knowledge of genetic mechanisms at the molecular level.

In 1976, we began the process of reading genetic information, at the rate of a few thousand 'letters' of genetic code per day. By 1995, we developed this technology to the point of reading millions of letters of genetic code per day, and began putting copies of the complete genetic code of entire organisms (an organism's 'genome') into computer data bases. The first complete genome sequenced was from the bacterium *Haemophilus influenza*, with 1.8 million genetic letters and about 1,700 genes. A yeast genome, sequenced in 1996, had 12 million genetic letters and about 6,000 genes. A nematode worm genome, sequenced in 1998, contained 97 million letters of genetic code and some 19,000 genes. A fruitfly genome, finished in 2000, had 180 million letters of genetic code and about 13,000 genes. The first plant genome sequenced, from the mustard family, also in 2000, contained 100 million letters of code and about 25,000 genes.

Two nearly complete copies of a human genome were assembled in computer databases in the year 2000, each representing about three billion letters of genetic code, and approximately 30,000 genes identified so far. Apparently, only about 2% of the human genome directly codes for protein synthesis, and the other 98% either has a different function, or possibly no direct function in human development. A mouse genome project, nearly complete, also shows about 3 billion letters of genetic code. Roughly half the genes in yeast, fruitflies and nematode roundworms match up with human genes, and so do many bacterial genes. A much higher proportion of human genes are shared in common with those of mice. Over 99.9% of the genetic information in the human genome, we now estimate, is identical among any two humans.

Since the 1960's, when a biological Rosetta stone was pieced together that allows us to translate this raw genetic language into the amino acid sequence that defines proteins, we've known how to read the simplest features of genetic maps directly from the DNA. It's only recently, however, within the last five years, that we've been able to see the entire set of genetic information for any cell or organism.

Maps of DNA sequences, their transcription into copies of genes and their translation into amino acid sequences that form proteins are the first levels of genetic information processing in living systems. By studying the structure, behavior and functions of molecules produced in cells according to genetic instructions, and by observing the consequences of mutations in DNA that alter the genetic information, we can follow and map out more complex pathways relating genetic information to biological characteristics of organisms.

LEFT

Molecular Genetics Laboratory Workbench
c. 1976–1990

Including the following:

Leonardo da Vinci drawing and map of gene expression throughout the human body

Microarray Anaylis in Progress,
data from Dr. Joanne Conover, Jackson Laboratory, and Rachel Allen, class of 2003

Yeast Protein Interactions,
contributed by Dr. Leroy Hood, Director, Institute of Systems Biology, Seattle, Washington

Scanner FMBIO II Multi-View
c. 1995
Courtesy of the New York State Police Forensic Investigation Center

Our new ability to analyze the entire set of genetic information in a cell, rather than just one gene at a time has given rise to new biological disciplines such as genomics and bioinformatics, proteomics, and computational biology. The ability to read an entire DNA sequence for a cell or organism, however, is still just the beginning of the genetic mapping project begun in 1911 by *ALFRED STURTEVANT*, who mapped six mutant traits in fruitflies to six points on their 'X' chromosome. Even though, today, we know the genetic map location of some 6,000 yeast genes, 13,000 fruitfly genes and 30,000 human genes, these maps currently consist of the simple sequence of genetic code letters, and in most cases little more than the amino acid sequence of proteins coded in the DNA.

One of the first new tools for mapping entire sets of genetic information into complex biological functions is the use of 'microarray' methods. Given that DNA is small enough so that a complete set of 3 billion letters of human DNA, for example, fits inside the nucleus of every cell in our body (actually two sets, since the mother and father each contribute a complete set), it is possible to place copies of thousands of different genes in a pattern on a glass slide, plastic film or other surface, using various techniques ranging from microchip fabrication technology to modified ink-jet printers. Microscopic sets of thousands of genes, arrayed in a pattern so that each gene is in a known location, can fit on a surface the size of a postage stamp. This 'microarray' can then be used as a probe, to determine how any of those genes are being expressed in a particular cell.

This technology is being used, for example, to determine which genes from the complete set in every cell, are being expressed in tumor cells versus non-tumor cells, in brain cells versus liver cells versus skin cells etc., to detect patterns of changes in gene expression during aging, and in thousands of other ways. This technique, which uses an initial map of the genome as a source of DNA probes and as a map reference point, allows us to analyze and compare whole sets of genetic information that are present in any organism or cells under any experimental conditions.

The ability to analyze the total set of genetic information in the cells of different organisms, and to associate subsets of genetic information with specific biological functions, is already transforming biology into a more unified science, since everything of biological and evolutionary relevance is, directly or indirectly a function of genetic information. We have, even at this early stage, recognized that some basic genetic information is shared in common among all organisms, and large branches of life's family tree share much more.

Genetic analysis at the genomic level is also transforming biology into more of an information science, where we can begin to understand the differences between living and non-living systems, and the variation and history of living systems, through patterns of information that are written, stored, copied, and expressed in unique ways characteristic of each kind of organism and cell.

It is clear from even the most basic understanding of genetics that genes only code directly for the sequence of molecules in RNA and the sequence of amino acids in protein, and therefore they can't code directly for more complex traits. On the other hand, it's also clear that very simple changes in the DNA sequence can cause profound and specific differences in the most complex features of an organism, including its behavior. Genetic maps begin with DNA sequences, and extend as far as we can follow the causal networks through molecular, cellular, developmental, anatomical, physiological, behavioral and evolutionary mechanisms. Limitations of our methods for genetic analysis, which until now have consisted mostly of analyzing one gene and one gene product at a time at the molecular level, probably underestimated the potential for coordinated patterns of gene expression involving many genes and gene products to orchestrate very complex biological functions. One indication that this is the case is the classical ability of plant and animal breeders, and scientists using laboratory populations, to rapidly modify the heredity of complex traits using selective breeding.

Gregor Mendel, Thomas Hunt Morgan, *ALFRED STURTEVANT*, and *WATSON* and *CRICK* in genetics are early pioneers in the field of genetic mapping. In the 136 years since Gregor Mendel recognized patterns of hereditary variation in pea plants that revealed the existence of genes, we've discovered DNA, the universal genetic code, and now we have begun to read and interpret all of the genetic information in living cells.

The late biologist Colin Pittendrigh, known for his contributions to the field of biological clocks, was fond of distinguishing non-living from living systems by describing the former as 'ordered,' and the latter as 'organized.' This leads to a simple definition of life as 'a system organized to reproduce itself.' The essential information that organizes living systems for the function of reproduction is the genetic information stored in DNA.

The analysis of information that occurs only in living systems is, ultimately, what distinguishes biological science from the other natural sciences. The hereditary genetic information stored in DNA connects the diverse forms and functions of life together through a common history of evolutionary adaptations. The science of genetics represents a unifying theoretical and material framework unique to the study of life. The knowledge we are seeking now in biology is the information needed for the organization of living systems that explains their unique functional properties. We will learn what the relationships are between genetic information and biological form and function by constructing maps relating one to the others, and then we will have a much more complete understanding of biology according to the newest and most exact observations. ⊕

BELOW
Nature
Vol. 377, no. 65475
(September 28, 1995)

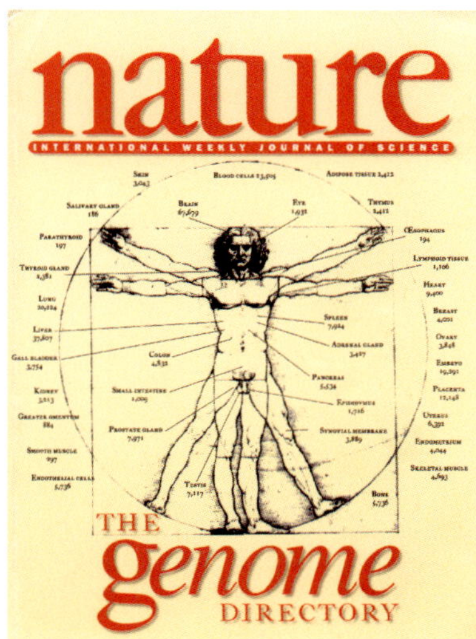

RICHARD LONG

When a map is made, three-dimensional objects are often reduced to a series of contours, outlines, and flat shapes. By using this process of abstraction, Richard Long transforms his actions into a sort of virtual map, often incorporating his walks into his work. He first used his hikes to make art in 1967, and has since become a leading figure in the evolution of conceptual art. During his walks, for example, he may document the lines his feet make in the grass, or travel according to road connections that coincide with geometric forms on a map. Long often presents texts and photographs to document his interactions with the landscape.

ABOVE

Stones in Ireland

1975

JOHN MCQUEEN

John McQueen combines traditional basket-making techniques with conceptual art processes to create organic sculptures inspired by the natural world. By changing the characteristics and scale of these two bodies of water, Lake Michigan and the Nile River, form and depth become easier to visualize. McQueen's subjects are constructed of sewn tree bark and scraped dried leaves. The sculpture's human scale suggests a surreal morphing from water body to organ system – a network in which we are carefully woven into our surroundings.

LEFT
Lake Michigan
1992

OPPOSITE
Up the Nile
1995

MATTHEW RITCHIE

With a dizzyingly elaborate cast of characters and a potent mix of mythology and science, Matthew Ritchie tackles the big bang and the emergence of life on earth. Into the vastness of such explorations he infuses a dose of soap opera, comic book action, and historical art references. Each elemental force is given a personality and background story with colors representing different players. Ritchie's work is sprawling and overwhelming – much like his subject matter. In the end his paintings aren't meant to act as an explanatory key of anything but his story, a story that maps the vastness of the universe.

OPPOSITE
Prelapse
1995

LORDY RODRIGUEZ

Inspired by his own fascination with maps, Lordy Rodriguez has drawn a dynamic new map of the United States using his imagination and life experience instead of traditional cartographic techniques of land surveys and satellite photographs. Rodriguez has reorganized the United States according to his desires and experiences. In his new country, the states are rearranged and take on new properties. Utah, for instance, is placed on the eastern coastline and other cities that actually lie on opposite ends of the country are juxtaposed side by side. Such relocations, geographic fantasies and assumptions about the United States add up to an individual, 21st century reinterpretation of the country.

RIGHT
New York
2000

OPPOSITE
Wyoming
2000

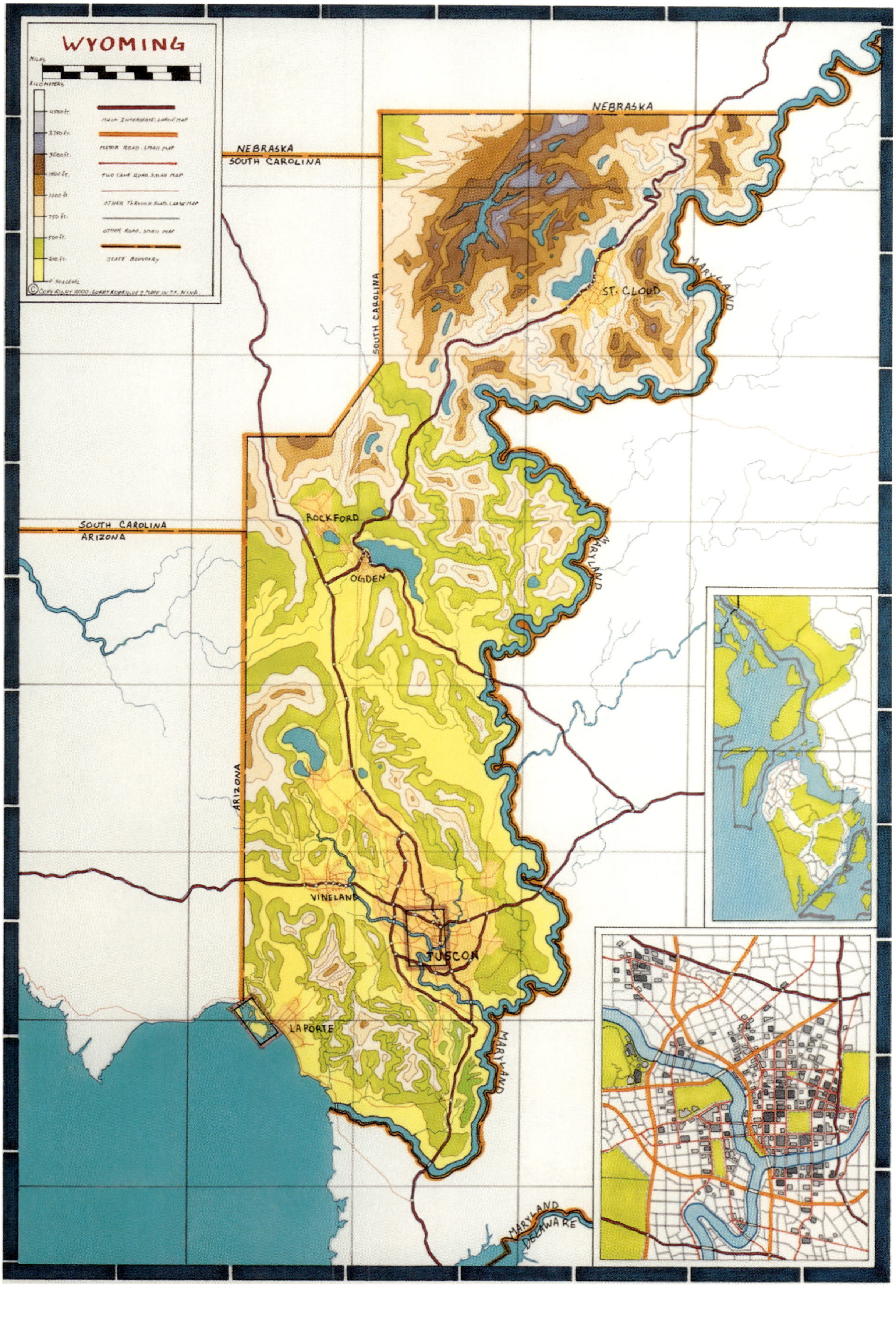
WYOMING
MILES
KILOMETERS
4500 ft.
3750 ft.
3000 ft.
1500 ft.
1000 ft.
750 ft.
500 ft.
200 ft.
MAIN INTERSTATE, LARGE MAP
MAJOR ROAD, SMALL MAP
TWO LANE ROAD, SMALL MAP
OTHER THROUGH ROAD, LARGE MAP
OTHER ROAD, SMALL MAP
STATE BOUNDARY
NEBRASKA
NEBRASKA
SOUTH CAROLINA
SOUTH CAROLINA
ST. CLOUD
MARYLAND
SOUTH CAROLINA
ARIZONA
ROCKFORD
OGDEN
MARYLAND
ARIZONA
VINELAND
TUSCON
LAPORTE
MARYLAND
MARYLAND
DELAWARE

ED RUSCHA

Ed Ruscha's pioneering artist books and iconic paintings continue to have great influence on contemporary art. In his bookwork, he reverses the traditional layout; using text on the cover and letting a series of images in the interior create a compelling narrative sequence. Ordinary places acquire the allure of legends. In the work, *Every Building on the Sunset Strip*, a long, continuous photograph of each building along this famous street creates a methodical map of Californian culture. Single-level storefronts and unlit neon signs capture the time period. As we scan from address to address we become the drivers cruising down the boulevard slowly observing a spread-out city.

OPPOSITE

Nine Swimming Pools
1968

Real Estate Opportunities
1970

Every Building on the Sunset Strip
1966

ERIKA AYALA STEFANUTTI

Erika Ayala Stefanutti uses metal work and jewelry design to address social issues, such as vanity, international commerce, and the hierarchy embedded within class structures. The latitude and longitude numbers that form the numerals of her necklaces correspond to the mines where the metal originates, and in this way they produce a map that is made from the place it describes. The stakes and politics that are intertwined with extracting a particular precious or semiprecious metal, and where it came from, are rarely considered by the person who wears it.

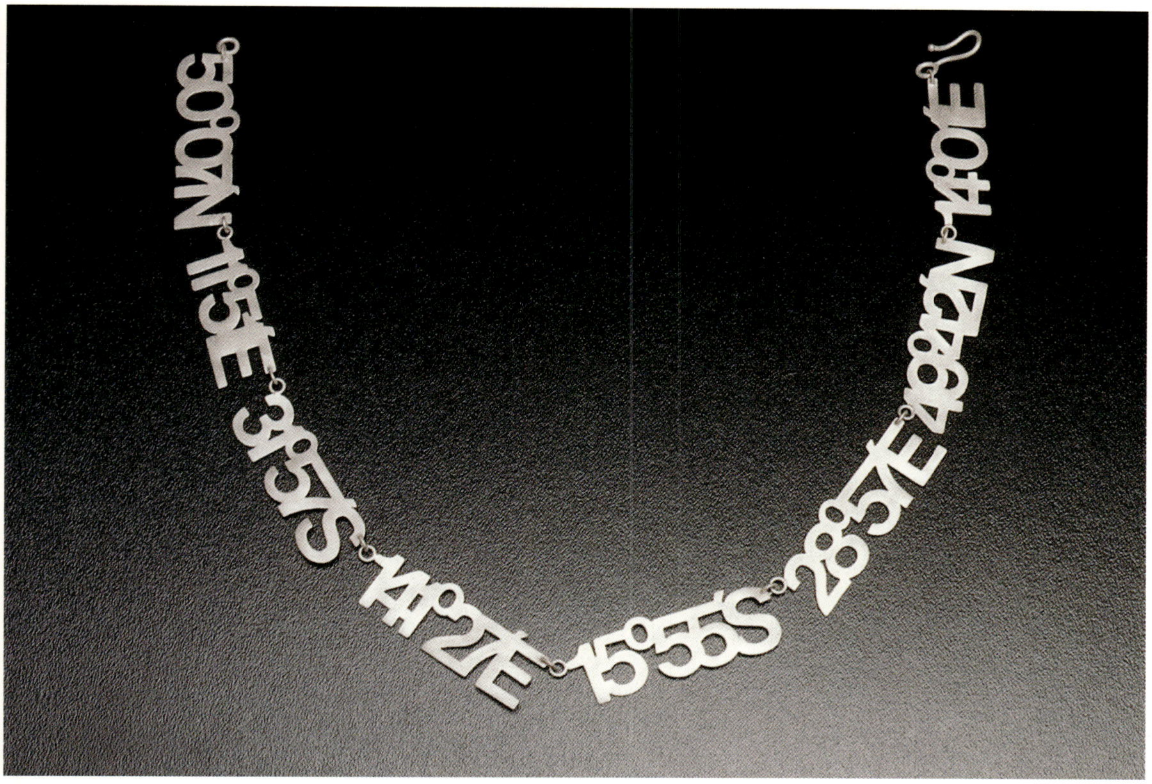

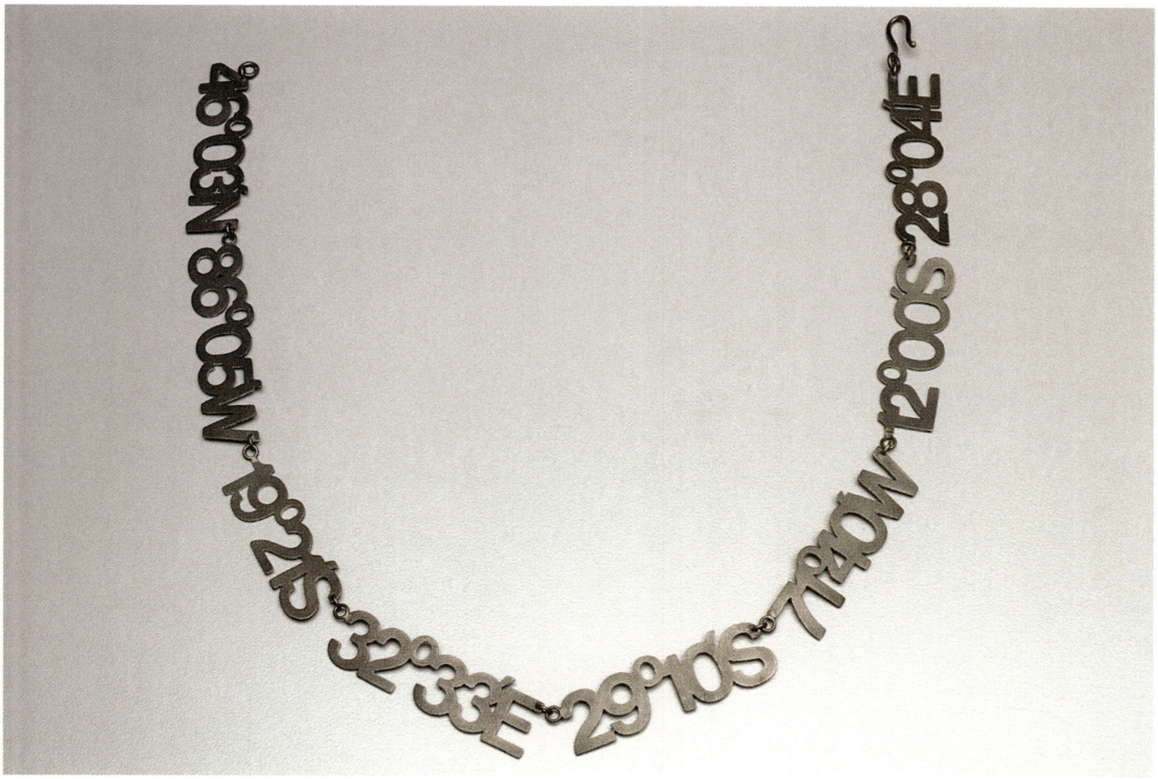

TOP
Coordinate Necklace: Silver Mine Sites
1996

BOTTOM
Coordinate Necklace: Copper Mine Sites
1996

ABOVE
The Haystack Radome Mapped with Its Own Telescope
2000

WENYON + GAMBLE

Wenyon and Gamble produced art that deals with the way we collect, store, and obtain knowledge. They combine backgrounds in art and science with cutting-edge technology, such as holography and digital photography. This sixteen-foot-long image was created by digitally combining frames taken from the radome of MIT's Haystack telescope in Westford, Massachusetts. The artists spent three years in residence at the observatory creating new work such as this print, which includes the technology, character and ambitions of one of the greatest map-making instruments on the planet.

MARC WISE

Marc Wise developed his CD-ROM as a way of mapping his travels through the New York City subway system. Wise gives personal dimension to the common subway map by capturing his experience of each stop on video. As he leaves the metro car at each station, Wise walks up to the street and records a 360-degree view. He adds his thoughts about each neighborhood and documents the food he's eaten during the days he took to complete the project. This diary-like journey through the dark tunnels of the subway system parallels food travelling through his body. The result is a narrative map, chronicling the quiet interaction of one individual within a densely populated urban space.

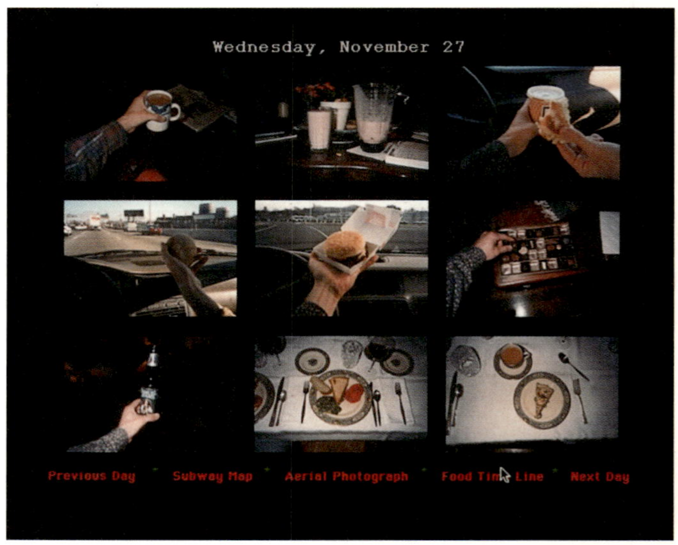

RIGHT & OPPOSITE
The Entire New York City Subway System and Everything I Ate for Seventy-Nine Days
1998

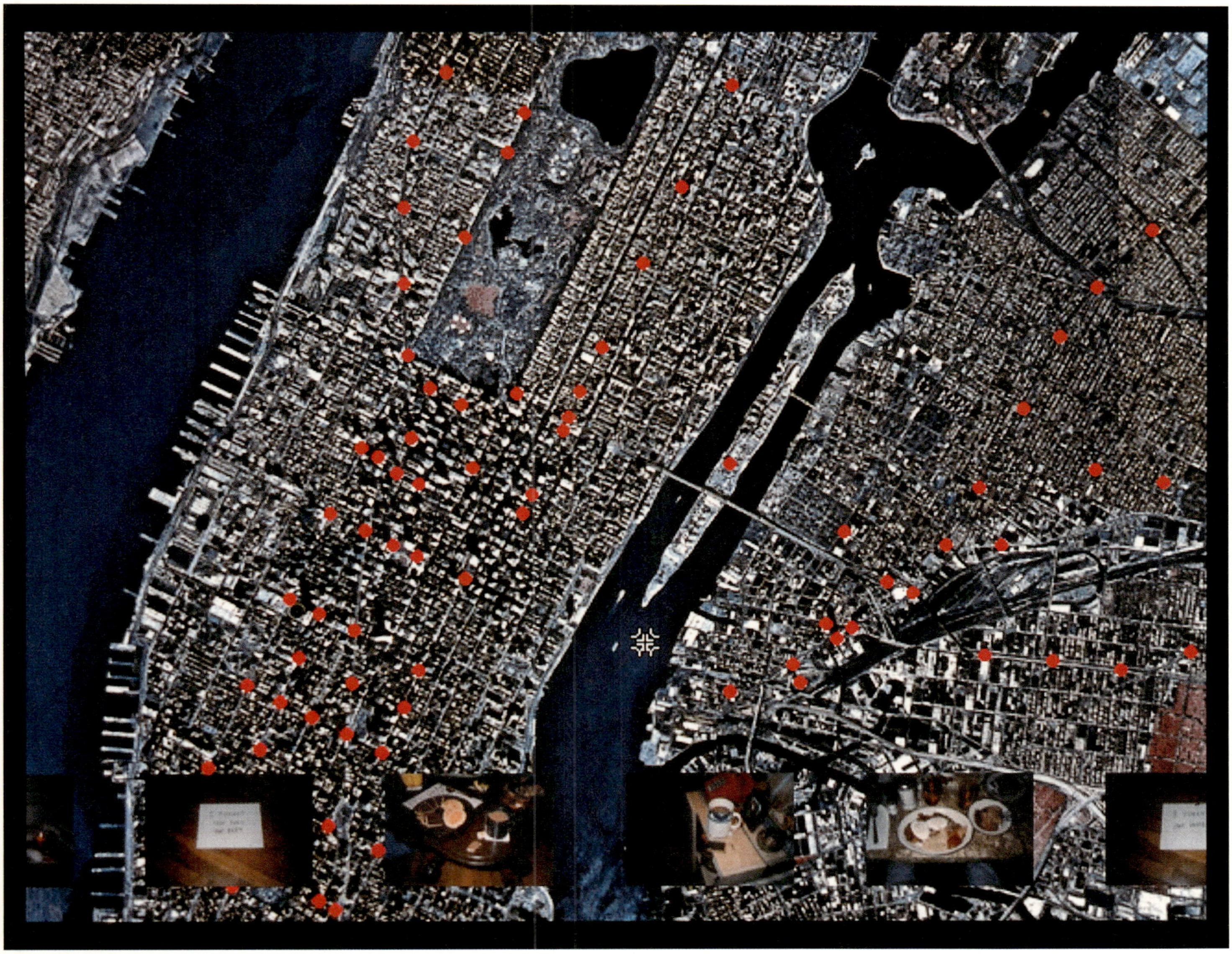

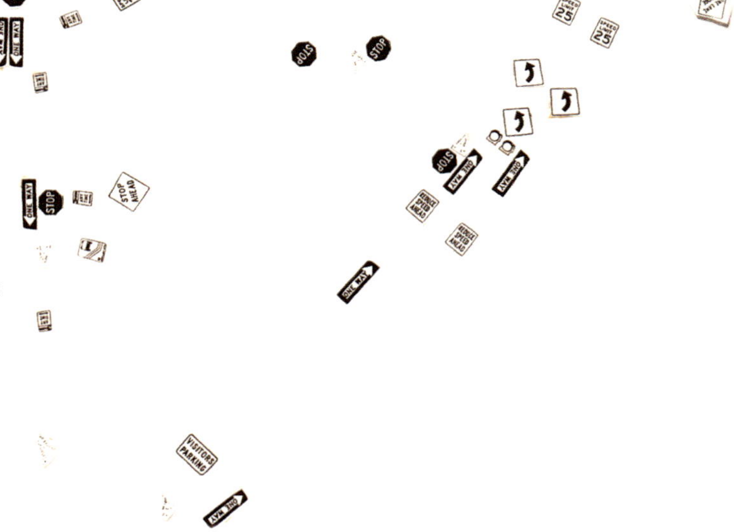

DENIS WOOD

What is important to know about your surroundings? Denis Wood has set out to understand and map the many ways he can know his neighborhood of Boylan Heights. What began as a design assignment for students at the School of Design at North Carolina State University has become a long-term project. Drawing from extensive experience in cartography and geography, Wood explores different views from which to map parts of his community. While many maps are created with a sense of detachment from the area they represent, Wood illustrates the character of his neighborhood by plotting its most intimate details, such as pumpkins found on Halloween night or graffiti frozen in the sidewalk.

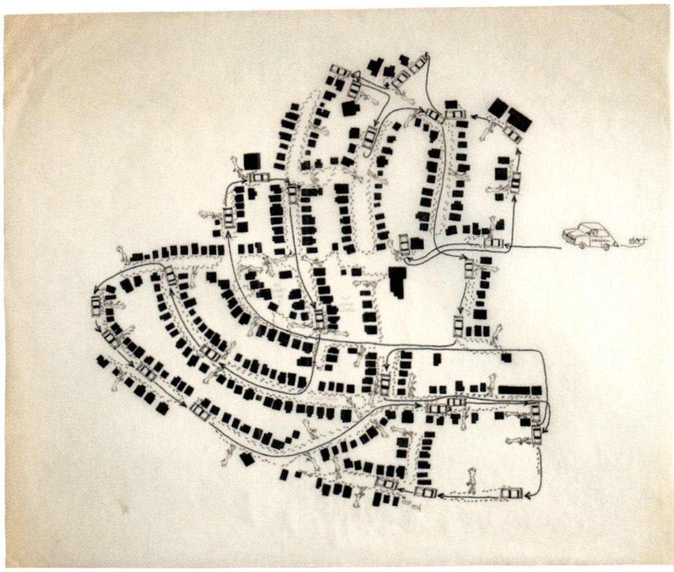

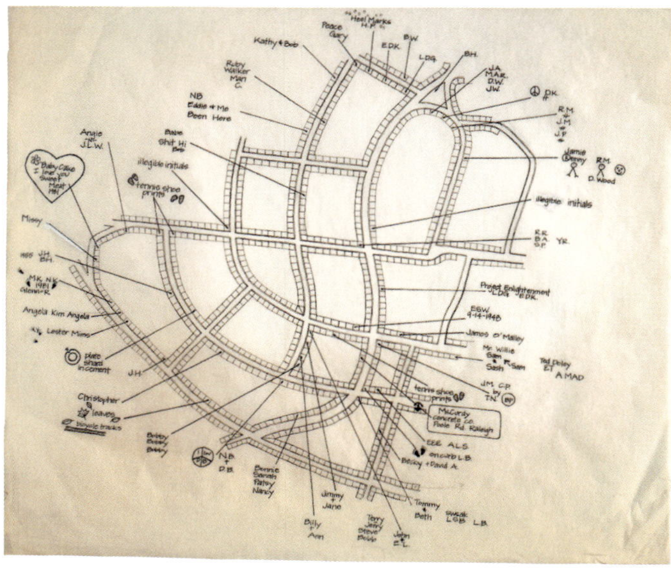

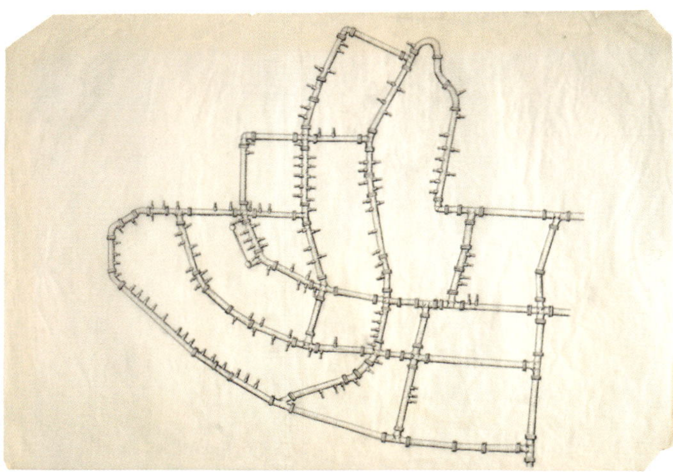

All drawings from ***Dancing and Singing: A Narrative Atlas of Boylan Heights*** *1982*

TOP & MIDDLE
by HELEN WALDROP and SUSAN EDWARDS

BOTTOM
by DAWN DAVIS and TIM HESS

OPPOSITE
by DENIS WOOD and CARTER CRAWFORD

T AS A WALL
ORTIONALLY
BLACK TYPE
LIFE LIVED
AINING
REPRESENTING
BASED ON
FE EXPECTANCY

MARE
DEL
NORT
MARE
DEL SUD
DEL
PERU

SELF-
DIVID
BETW
REPR
AND T
WHIT
LIFE
STAT

CONTRIBUTORS

SUSAN BENDER is associate dean of the faculty and associate professor of anthropology at Skidmore College. She received her MA and PhD at the State University of New York at Albany. Her field work includes research at the Jackson Hole Archaeological Research Project, the Fish Creek Archaeological Project, the 18th Century Sword's farmstead in Schuylerville and a site survey at the confluence of the Battenkill and Hudson River. The many courses she teaches at Skidmore include *Anthropology of the Human Past*, *Mesoamerican Archaeology*, *Analysis of the Human Skeleton*, and *Women and Gender in Evolutionary Perspective*.

IAN BERRY is curator of the Tang Teaching Museum and Art Gallery at Skidmore College. A specialist in contemporary art, he received his MA in Curatorial Studies at the Center for Curatorial Studies at Bard College. As assistant curator at the Williams College Museum of Art, he organized several exhibitions including solo presentations of James Casebere, Jane Masters, William Wegman, and William B. Schade. Berry was guest curator of the inaugural exhibition *Showroom* and the exhibition, *Writings: Jenny Holzer and William Kennedy* for the Arts Center of the Capital Region in Troy, NY. His recent projects include *Chain Reaction: Rube Goldberg and Contemporary Art*, *Work: Shaker Design and Recent Art* and the first exhibition in the Tang's Opener series *Jonathan Seliger: Floor Model.*

BERNARD POSSIDENTE received his BA in Biology at Wesleyan University in 1976 and his PhD in Genetics at the University of Iowa in 1981. He joined the faculty of Skidmore College in 1983 and served as chair of the Biology department from 1990–1994. Currently professor of biology at Skidmore, Possidente has been a collaborator and principal investigator on many projects and teaches courses in genetics, human genetics, history of genetics and on biological clocks. His research centers on the study of quantitative genetic analysis of biological clock function in mouse and fruitfly model systems and the genetic analysis of behavioral adaptations.

RICHARD WILKINSON is professor of biological anthropology at the University at Albany. He received his PhD from the University of Michigan in 1970 and his research interests are in the areas of bioarchaeology, osteology and forensic anthropology as he is interested in elucidating population processes, behaviors and interactions among groups for whom there is no written record. To this end he has authored many papers including those on prehistoric cranial surgery, violence-induced trauma, population movements and demographic structures. Dr. Wilkinson's teaching interests include functional skeletal anatomy, human evolution, primate behavior and population dynamics of prehistoric and historic human groups.

1925 1930 1935 1940 1945 1950 1955

DAVID BAKALAR ◉ Born in 1924, Chestnut Hill, Massachusetts

DB • MS, Harvard University, Cambridge, Massachusetts
DB • BS, Harvard College, Cambridge, Massachusetts
DB • ScD, MIT, Cambridge, Massachusetts

CAROL BARTON ◉ Born in 1954, St. Louis, Misso

KATE ERICSON ◉ Born in 1955, New Yor
MEL ZIEGLER ◉ Born in 1956, Cam

JOYCE KOZLOFF ◉ Born in 1942, Somerville, New Jersey

RICHARD LONG ◉ Born in 1945, Briston, England

JOHN MCQUEEN ◉ Born in 1943, Oakland, Illinois

ED RUSCHA ◉ Born in 1937, Omaha, Nebraska

MICHAEL WENYON ◉ Born in 1955, Dayton,
SUSAN GAMBLE ◉ Born in 1957,

DENIS WOOD ◉ Born in 1945, Cleveland, Ohio

0 5
1 INCH = 5 YEARS

CB • *Maryland Biennial*, Works on Paper, Baltimore Museum of Art,

THÉODORE COULOMBE ◉ Born in 1963, Manhasset, New York

ALAN KRATHAUS ◉ Born in 1968, Albany, New York

WIM DELVOYE ◉ Born in 1965, Wervik, Belgium

SAM EASTERSON ◉ Born in 1972, Hartford, Connecticut

ew York

l town, Pennsylvania

KE • MFA, California Institute of the Arts, Valencia, California

MZ • MFA, Rhode Island School of Design, Providence, Rhode Island

KE • BFA, Kansas City Art Institute, Kansas City, Missouri

NINA KATCHADOURIAN ◉ Born in 1968, Stanford, California

JK • BFA, Carnegie Institute of Technology, Pittsburgh, Pennsylvania

JK • *Visionary Ornament*, Boston University Art Gallery,

JK • MFA, Columbia University, New York, New York

JK • *An Interior Decorated*, Tibor de Nagy Gallery, New York, New York

ICAH LEXIER ◉ Born in 1960, Winnipeg, Manitoba

ML • BFA, University of Manitoba, Winnipeg, Manitoba

ML • MFA, Nova Scotia College of Art & Design, Halifax, Nova

RL • *Artist of the Map: Image/Process/Data/Place*, Helen Foresman, Lawrence,

RL • *Projects*, The Museum of Modern Art, New York, New York

RL • The Solomon R.Guggenheim Museum, New York,

JMcQ • BA, University of South Florida, Tampa, Florida

JMcQ • *Fiber R/Evolution*, Milwaukee Art Museum, Milwaukee

JMcQ • MFA, Tyler School of Art, Temple University, Philadelphia, Pennsylvania

MATTHEW RITCHIE ◉ Born in 1964, London, England

MR • BFA, Camberwell School of Art, London

MR • Boston University, Boston, Massachusetts

LORDY RODRIGUEZ ◉ Born in 1976, Manila, Philippines

ER • Ferus Gallery, Los Angeles, California

ER • *The Works of Edward Ruscha*, San Francisco Museum of Modern Art,

ERIKA AYALA STEFANUTTI ◉ Born in 1965, Travers City, Michigan

hio

ondon, England

G • BA Fine Art, Goldsmiths' College, University of London, England

W • BSc, Bristol University

W • MSc, Imperial College, University of London, England

MARC F. WISE ◉ Born in 1964, Washington, DC

MW • BFA, Rhode Island School of Design, Provi-

DW • BA, Case-Western Reserve University, Cleveland, Ohio

DW • *Annual Wake County Artists Exhibition*, Raleigh, North Carolina

DW • *41st North Carolina Artists Exhibition*, North Carolina Museum of Art

DW • MA and PhD, Clark University, Worcester, Massachusetts

DW • *39th Traveling Exhibition of Works by Artists of the Western Reserve*, Cleveland Museum of Art, Cleveland, Ohio

1990 1991 1992 1993 1994 1995 1996

DB • *David Bakalar, Sculpture and Paintings*, The MIT Museum, Cambridge, Massachusetts
DB • *Icons for the 21st Century*, The William Benton Museum of Art, University of Connecticut, Storrs, Connecticut
DB • *The Avant Garde Heritage, Sculpture by David Bakalar*, The State Russian Museu
DB • Kouras Gallery, New York

Baltimore, Maryland
CB • *A World of Maps*, Anchorage Museum of History & Culture, Anchorage, Alas
CB • *Artists' Books*, Tate Gallery, London, England

AK • BA, State University of New York at Plattsburgh, Plattsburgh, New York
AK • MFA, Cranbrook Academy of Art, Bloomfield Hills, Michigan
TC+AK • *Theater of Operations*, Atlanta College of Art Gallery, Atla

WD • *Cartographic Paintings*, Galleria Andrea Murnik, Milano
WD • *Dokumenta IX*, Kassel
WD • Sonnabend Gallery, New York

SE • BFA, The Cooper Union for the Advancement of Science and Art, New York, N

KE+MZ • *Dark on that Whiteness*, Wolff Gallery, New York, New York
KE+MZ • *Places with a Past: Site Specific Art in Charleston*, Spoleto Festival, Charleston, South Carolina
KE+MZ • James Hockey Gallery, WSCAD, Farnham, England
KE+MZ • *Mapping: A Response to MoMA*, American Fine Arts, C
KE+MZ • *The Body As Measure*, Davis Museum, Wellesley Colle
KE ◉ Died in 1995

NK • BA, Brown University, Povidence, Rhode Island
NK • MFA, University of California, San Diego, San Diego, California
NK • *Carpark*, inSITE 94, San Diego, California
NK • Whitney Museum Independent St

Boston, Massachusetts

ML • Art Gallery of Ontario. Toronto, Ontario
Scotia
ML • *The Body as a Measure*, Davis Museum and Cultural Center, Wellesley, Massachus
ML • McIntosh Gallery, The University of Western Ontario, London, Ontario

Kansas
RL • The Tate Gallery, London
RL • *Richard Long: Here and There*, Mo

New York

JMcQ • 1988 *Knots and Nets*, Herbert F. Johnson Museum, Ithaca, New York
Wisconsin
JMcQ • *The Language of Containment*, Renwick Gallery, Museum of American Art, Washington, D.C.

MR • *Working Model*, Basilico Fine Arts, New York, New Y

San Francisco, California

ES • BFA, The University of Michigan, Ann Arbor, Michigan
ES • MFA, Cranbrook Academy of Art, Bloomfield Hills, Michigan
ES • *Thirteen X's and One Y*, Gallery 414, Ann Arbor, Michigan
ES • *Signals: Late Twentieth Century*
ES • *Viewpoint: Art as Message*, Craft

W+G • *Bibliography*, Art Tower Mito, Mito, Japan
G • Fellow, Center for Advanced Visual Studies, MIT, Cambri
W+G • *Volumes*, The Photographer's Gallery, London, England and Collins Gallery, Glasgow
W+G • *Light & Dark*, Royal Observatory, Edinburgh, Scotland

MW • Opsis Foundation, New York
MW • *On the Road*, Museum of Fine Arts, Houston, Texas
dence, Rhode Island
MW • Center for Creative Studies, Detroit, Michigan

0 1
1 INCH = 1 YEAR

1997 1998 1999 2000 2001

DB ◉ Lives and works in Chestnut Hill, Massachusetts

Petersburg, Russia

CB ◉ Lives and works in Glen Echo, Maryland

CB • *Artists on the Road: Travel*, National Museum of Women in the Arts, Washington D.C.

CB • *The Art of the Book: Visualizing Poetry*, Swarthmore College, Swarthmore, Pennsylvania

TC+AK • Group Show, 57 Hope St. Gallery, Brooklyn, New York

TC ◉ Lives and works in Brooklyn, New York

AK ◉ Lives and works in Houston, Texas

rgia

TC+AK • *Playing Off Time*, Aldrich Museum of Contemporary Art, Ridgefield, Connecticut

TC+AK • *Showroom*, The Arts Center of the Capitol Region, Troy, New York

WD • *Cartographers*, Museum of Contemporary Art, Zagreb, Croatia

WD ◉ Lives and works in Belgium

WD • *48e Biennale di Venezia, d'AperTutto*, Venice, Italy

k

SE • MS, University of Minnesota, Minneapolis, Minnesota

SE ◉ Lives and works in North Hollywood, California

SE • *Biennial Exhibition*, Whitney Museum of American Art, New York, New York

SE • RARE Gallery, New York; Art in General, New York

SE • *Paradise 8*, Exit Art/The First World, New York

SE • *Dialogues: Sam Easterson/T.J. Wilcox*, Walker Art Center, Minneapolis, Minnesota

York, New York;
lesley, Massachusetts

MZ ◉ Lives and works in Austin, Texas

gram

NK • *Cartographers*, Museum of Contemporary Art, Zagreb, Croatia

NK ◉ Lives and works in Brooklyn, New York

NK • *Rewriting the City*, Bard College, Annandale-on-Hudson, New York

NK • *Maps, Charts, and Routes*, Irvine Center for the Arts, Irvine, California

NK • *Mended Spiderwebs and Other Natural Misunderstandings*, Debs & Co., New York, New York

JK ◉ Lives and works in New York, New York

JK • *Crossed Purposes*, Mills College Art Museum, Oakland, Califonia

JK • *Targets*, DC Moore Gallery, New York, New York

JK • *Other People's Fantasies: Maps, Movies, and Menus*, DC Moore Gallery, New York, New York

ML ◉ Lives and works in New York, New York

ML • Musee d'art Contemporian de Montreal, Montreal, Quebec

ML • *The Time of Our Lives*, New Museum of Contemporary Art, New York, New York

Museum of Fort Worth, Fort Worth, Texas

RL ◉ Lives and works in Bristol, England

JMcQ ◉ Lives and works in Saratoga Springs, New York

JMcQ • *Boundaries*, Elliott-Brown Gallery, Seattle, Washington

JMcQ • *Weaving the World Art of Linear Construction*, Yokohama Museum of Art, Yokohama, Japan

MR ◉ Lives and works in New York, New York

MR • *Biennial Exhibition*, Whitney Museum of American Art, New York, New York

MR • *Concentrations 38: The Slow Tide*, The Dallas Museum of Art, Dallas, Texas

MR • *The Big Story*, Cleveland Center for Contemporary Art, Cleveland, Ohio

MR • *Map the Gap*, Storefront of Art & Architecture, New York, New York

LR • University of Houston, Houston, Texas

LR • BFA, School of Visual Arts, New York, New York

LR ◉ Lives and works in Houston, Texas

LR • *The City*, Nicole Klagsbrun Gallery, New York, New York

LR • *All Sizes Real People*, Museum of Fine Arts, Houston, Texas

LR • *States*, Clementine Gallery, New York, New York

LR • *At Home and Abroad: 20 Filipino Artists*, Asian Art Museum, San Francisco, California

ER ◉ Lives and works in Venice, California and the Mojave Desert

ER • *Edward Ruscha: Editions 1959–1999*, Walker Art Center, Minneapolis, Minnesota

ER • *Biennial Exhibition*, Whitney Museum of American Art, New York, New York

ER • *Landmark Pictures*, Busch-Reisinger Museum, Harvard University, Cambridge, Massachusetts

n Jewlery, Cranbrook Art Museum, Bloomfield Hills, Michigan

ES ◉ Lives and works in Bloomfield, Michigan

ES • *Off the Map*, Paint Creek Center for the Arts, Rochester, Michigan

, Center for the Visual Arts, St. Louis, Missouri

ES • *The Commerative Cup*, Castle Gallery, College of New Rochelle, New Rochelle, New York

sachusetts

W+G ◉ Lives and works in Jackson Heights, New York

G • Mphil History and Philosophy of Science, Cambridge University, Cambridge, Massachusetts

W+G • *Bibliomancy*, The Boston Anthaeum, Boston, Massachusetts

W+G • *Observing the Observers*, Compton Gallery, MIT Museum, Cambridge, Massachusetts

MW ◉ Lives and works in Brooklyn, New York

MW • MFA, School of Visual Arts, New York

MW • *Portrait of the Force*, a photography/video project for the New York City Police Museum, New York, New York

DW ◉ Lives and works in Raleigh, North Carolina

CHECKLIST

Dimensions are in inches, height precedes width precedes depth

BERNHARD SIEGRFIED ALBINUS
TABULAE SCELETI ET MUSCULORUM CORPORIS HUMANI, c. 1747
30 x 48
Collection of Malloch Rare Book Room, New York Academy of Medicine

DAVID BAKALAR
WHEEL OF FORTUNE, 1988
Anodized aluminum with scotchal tape
20 x 13.5 diameter
Courtesy of the artist

CAROL BARTON
TUNNEL MAP, 1988
Silkscreen/offset edition
7.75 x 7.75 x 10 (extended)
Special Collections of Lucy Scribner Library, Skidmore College

EVERYDAY ROAD SIGNS, 1988
Silkscreen/offset edition
7.25 x 8.75 x 18 (extended)
Courtesy of the artist

JEAN-BAPTISTE MARC BOURGERY
VOLUME 6 OF TRAITÉ COMPLET DE L'ANATOMIE DE L'HOMME, 1852
Hand-colored lithograph
17.125 x 13
Collection of Malloch Rare Book Room, New York Academy of Medicine

L. R. BURLEIGH
SARATOGA SPRINGS, NY, 1888
25 x 38
Collection of the Geography and Map Division, Library of Congress, Washington, DC

DAVID H. BURR/ DISTURNELL
HUDSON RIVER & VICINITY, 1835
4 x 28
Manuscripts and Special Collections, New York State Library, Albany

CHAMBER OF COMMERCE
FUN AND REST... BE SARATOGA'S GUEST. THE "VACATION" CONVENTION CITY, 1952
15.5 x 16
Collection of the Geography and Map Division, Library of Congress, Washington, DC

ARNOLD COLOM
PASCAARTE VAN NIEU NEDERLANDT UYTGGEVEN DOOR HENDRICK DONCKER, 1656
23.5 x 21
Courtesy of the Voorhees Collection Alexandria, Virginia

ASTLEY PASTON COOPER
ON THE ANATOMY OF THE BREAST, 1840
14.75 x 10.75
Collection of Malloch Rare Book Room, New York Academy of Medicine

J.A. COREY
MEMORY MAP, 1936
13 x 7.75
Courtesy of the Historical Society of Saratoga Springs

THÉODORE COULOMBE and ALAN KRATHAUS
AIR TO GROUND TRIANGULATION. BRIDGEHAMPTON, N.Y. E.T.O U.S.A. 12.13.94
From *Signal Corps Series*, 1994
Cibachrome Print
36 x 48
Courtesy of the artists

BEACH HEAD. BRIDGEHAMPTON, N.Y. E.T.O. U.S.A 12.13.94
From *Signal Corps Series*, 1994
Cibachrome Print
48 x 48
Courtesy of the artists

WIM DELVOYE
SOUTH VIKXONIA, 1992
Silkscreen ink on PVC
82.5 x 104
On extended loan to The Tang Teaching Museum and Art Gallery from a Private Collection

SAM EASTERSON
ANIMAL, VEGETABLE, VIDEO: MEMORIES OF MANHATTAN FROM A MILLENNIUM AGO, 2000
(Project was made during an Artist Residency at Art in General, New York)
Mixed media
Dimensions variable
Accompanying video: 5:04 minutes
Courtesy of the artist and RARE Gallery, New York

KATE ERICSON and MEL ZIEGLER
DARK ON THAT WHITENESS, 1987
Latex paint, sand-blasted glass, steel shelves
147 x 192 x 3.5
Courtesy of the artist

J. D. FISK
NEW YORK, SARATOGA SPRINGS (CITY), 1968
30.25 x 20.25
Collection of the Geography and Map Division, Library of Congress, Washington, DC

VENICE FRISIUS/ APIAN CHARTES
COSMOGRAPHIQUES AVEC LES NOMS PROPRIETEZ, NATURE & OPERATIONS DES VENTS, 1581
9.5 x 14
Courtesy of the Voorhees Collection Alexandria, Virginia

NINA KATCHADOURIAN
HANDHELD SUBWAY, 1996
Cibachrome
16 x 20
Courtesy of the artist and Debs & Co. Gallery

AUSTRIA, 1997
Excised paper map
7 x 5
Courtesy of the artist and Debs & Co. Gallery

FINLAND'S LONGEST ROAD, 1999
Excised paper map in petri dish
5.5 diameter
Courtesy of the artist

MAP DISSECTION I, 1991–97
Excised paper map between glass
26 x 38
Collection of the artist

JOYCE KOZLOFF
KNOWLEDGE #75: 1290, 1999
Gesso and watercolor on cardboard with wood and plastic base
32" circumference
Courtesy of the artist and DC Moore Gallery

KNOWLEDGE #78: 1154, 2000
Watercolor, acrylic, plaster, and rope on cardboard
9.125 diameter
Courtesy of the artist and DC Moore Gallery

LA CONQUISTA, 1999
Watercolor, plaster, and rope on cardboard with wood and metal base
12.75 diameter
Courtesy of the artist and DC Moore Gallery

TARGETS, 2000
Acrylic on canvas with wood frame
108 diameter
Courtesy of the artist and DC Moore Gallery

MICAH LEXIER
SELF-PORTRAIT AS A WALL DIVIDED PROPORTIONALLY BETWEEN THIS BLACK TYPE REPRESENTING LIFE LIVED AND THE REMAINING WHITE SPACE REPRESENTING LIFE TO COME, BASED ON STATISTICAL LIFE EXPECTANCY, 1998
Vinyl type on existing wall
Dimensions based on installation
Courtesy of the artist and Jack Shainman Gallery, New York

RICHARD LONG
STONES IN IRELAND, 1975
Black and white photograph
17 x 24.5
On extended loan to The Tang Teaching Museum and Art Gallery from a Private Collection

GIOVANI GRIMANI
MARE DEL NORD, 17th century
17.75 x 23.5
Courtesy of the Voorhees Collection Alexandria, Virginia

HONDIUS
NOVA ANGLIA NOVUM BELGIUM ET VIRGINIA, 1638
18.5 x 20.75
Courtesy of the Voorhees Collection Alexandria, Virginia

INTERSTATE PUBLISHING COMPANY
MAP OF SARATOGA SPRINGS, NEW YORK, 1955
22.75 x 33.75
Collection of the Geography and Map Division, Library of Congress, Washington, DC

JAN JANSSON
AMERICAN SEPTENTRIONALIS AMSTELODAMI, 1635
20.5 x 23.75
Collection of the Geography and Map Division, Library of Congress, Washington, DC

JOHANNES DE KETHAM
FASCILIUS MEDICINAE IN QUO CONTINTENTUR: VIDELICT, c. 1495
12 x 8.5
Collection of Malloch Rare Book Room, New York Academy of Medicine

HENRY LAWRENCE REALTY
MAP OF SARATOGA SPRINGS AND EXCELSIOR PARK, 1881
13.75 x 11.75
Collection of the Geography and Map Division, Library of Congress, Washington, DC

L.H. LAWRENCE and SAMUEL J. MOTT
CITY PLANNING BOARD MAP, REVISION OF THE 1939 MAP, 1949
37 x 54
Collection of the Geography and Map Division, Library of Congress, Washington, DC

LESTER BROTHERS REALTY
DRIVES IN THE VICINITY OF SARATOGA SPRINGS, 1887
12.25 x 13
Collection of the Geography and Map Division, Library of Congress, Washington, DC

LOTTER/SEUTTER
RECENS EDITA TOTIUS NOVI BELGII IN AMERICA SEPTENTRIONALI..., 1757
20 x 23.25
Courtesy of the Voorhees Collection Alexandria, Virginia

MAPINFO SPECTRUM
INTERACTIVE WEBSITE, LAUNCHED FEBRUARY, 2001
www.mapinfo.com

JOHN McQUEEN
UP THE NILE, 1995
Bark, thread
85 x 40 x 18
Collection of Robert Pfannebecker

LAKE MICHIGAN, 1992
Bark, thread
44 x 24 x 10
Courtesy of the artist

MERRIAN AND MOORE
THE FRANKLIN IMPROVED CELESTIAL GLOBE, c. 1850
Collection of The New York State Museum, Albany

HERMAN MOLL
NEW AND EXACT MAP OF THE DOMINIONS OF THE KING OF GREAT BRITAIN ON YE CONTINENT OF NORTH AMERICA, 1735
40 x 26
Courtesy of the Voorhees Collection, Alexandria, Virginia

HENRY POPPLE
A MAP OF THE BRITISH EMPIRE IN AMERICA WITH THE FRENCH & SPANISH SETTLEMENTS ADJACENT THERETO, 1733
41 x 25.75
Courtesy of the Voorhees Collection, Alexandria, Virginia

MATTHEW RITCHIE
PRELAPSE, 1995
Oil and marker on canvas
90 x 84
On extended loan to The Tang Teaching Museum and Art Gallery from a Private Collection

JACOB ROBYN
LIGHTNING SEA-COLUMN, OR SEA-MIRROR: DISCOVERING ALL THE COASTS AND ISLANDS OF EUROFA, AFRICA, AMERICA, AND ASIA. WITH A PLAIN DESCRIPTION THEREOF;..., 1689
20 x 16
Manuscripts and Special Collections, New York State Library, Albany

LORDY RODRIGUEZ
UTAH, 1999
Ink on paper
17 x 36
Collection of Nick Debs

NEW YORK, 2000
Ink on paper
31 x 24
Collection of Eleanor Williams

WYOMING, 2000
Ink on paper
27 x 22
Private Collection, Detroit, Michigan

ED RUSCHA
NINE SWIMMING POOLS, 1968
Offset lithographic book, 64 pp, 7 x 5.5
Collection of The Tang Teaching Museum and Art Gallery
2000.69

REAL ESTATE OPPORTUNITIES, 1970
Offset lithographic book
Collection of The Tang Teaching Museum and Art Gallery
2000.70

EVERY BUILDING ON THE SUNSET STRIP, 1966
Offset lithographic book, 7 x 291.5
Collection of The Tang Teaching Museum and Art Gallery
2000.73

C.J. SAUTHIER
TOPOGRAPHICAL MAP OF THE PROVINCES OF NEW YORK AND NEW JERSEY, 1777
34 x 24.5
Manuscripts and Special Collections, New York State Library, Albany

JAMES SCOTT
PROPERTY OF GIDEON PUTNAM, 1810
12 x 23
Courtesy of the Historical Society of Saratoga Springs

LE SIEUR S.A. AMSTERDAM, CHEZ PIERRE MORTIER LIBRAIRE
CARTE NOUVELLE DE L'AMERIQUE ANGLOISE... CONTENANT (PARTIE ORIENTALE), c.1700
Courtesy of the Voorhees Collection, Alexandria, Virginia

JOHN SMITH
THE GENERALL HISTORIES OF VIRGINIA, NEW ENGLAND, AND THE SUMMER ISLES, 1632
11.25 x 7.5
Courtesy of the Voorhees Collection, Alexandria, Virginia

ERIKA AYALA STEFANUTTI
COORDINATE NECKLACE: SILVER MINE SITES, 1996
Fine silver
8 x 8 with base approx. 9 x 9 x 2
Courtesy of the artist

COORDINATE NECKLACE: COPPER MINE SITES, 1996
Copper
8 x 8 with base approx. 9 x 9 x 2
Courtesy of the artist

ALFRIDA STORM
SKIDMORE COLLEGE, 1939
20 x 25.5
The Tang Teaching Museum and Art Gallery

W.H. THOMPSPON
ARTERIAL SYSTEM IN MAN
25 x 4.5
Collection of The New York State Museum, Albany

LE SR. ROBERT DE VAUGONDY
PARTIE DI L'AMERIQUE SEPTENTRIONALE, QUI COMPREHEND LE DE L'OHIO..., 1755
19.5 x 25
Courtesy of the Voorhees Collection, Alexandria, Virginia

ANDREAS VESALIUS
SKELETON, POSTERIOR VIEW FROM DE HUMANI CORPORIS FABRICA, BOOK 1, 1543 (printed 1934)
Woodcut
13.5 x 9
Collection of Malloch Rare Book Room, New York Academy of Medicine

VISSCHER
NOVII BELGII NOVAEQUE ANGLIAE NEC NON PARTIS VIRGINIAE TABULA...., 1685
20.25 x 24
Courtesy of the Voorhees Collection, Alexandria, Virginia

F.L. WAGAN
HOTEL CONTINENTAL SARATOGA SPRINGS, NEW YORK, 1921
10.5 x 8
Collection of the Geography and Map Division, Library of Congress, Washington, DC

WENYON + GAMBLE
THE HAYSTACK RADOME MAPPED WITH ITS OWN TELESCOPE, 2000
Iris print on tyvek
40 x 200
Courtesy of the artists

MARC WISE
THE ENTIRE NEW YORK CITY SUBWAY SYSTEM AND EVERYTHING I ATE FOR SEVENTY-NINE DAYS, 1998
CD-ROM
Courtesy of the artist

DENIS WOOD
DRAWINGS AND PHOTOGRAPHS FROM DANCING AND SINGING: A NARRATIVE ATLAS OF BOYLAN HEIGHTS, 1982–89
Pen and pencil on paper
Courtesy of the artist

ABOUT FACES, 2000
CD-ROM
HT Medical Systems, Rockville, Maryland

ACUPUNCTURE HEAD, 2001
Soft vinyl
Anatomical Chart Company

AFFYMETRIX DNACHIP, c. 2000
Courtesy of Ari Hobfoll '00, Microarray and DNA Sequencing Facilty, Albert Einstein School of Medical at Yeshiva University, The Bronx, New York

DOUBLE HELIX MODEL, c. 1953
6 x 12 x 6
Courtesy of Dr. James Watson, Cold Spring Harbor Laboratory

EARLY 20TH CENTURY FRUITFLY GENETICS LABORATORY, c. 1913
Including the following reproductions
Homunculus, c. 1596, After Delenpatius;
Recipe for Mice, c. 1620, After Van Helmont;
Calvin Bridges, *Fruitfly Anatomy*, c. 1910, Drosophila Information Service Study Guide;
Research Notebook Page of Genetic Mapping Data, c. 1911–1913 by Alfred H. Stuyvesant, courtesy of California Institute of Technology Archives;
First Genetic Map, from original publication in Journal of Experimental Zoology, Volume 14, pages 43-59, 1913
Collection of the New York State Library

EUROPEAN MEDICAL MANNEQUINS, 18th centuries
Ivory, silk and wood
5 to 8
Collection of Malloch Rare Book Room, New York Academy of Medicine

THE FOUR SEASONS, 2000
Duke University
CD-ROM
Published by Duke University School of Medicine

THE HUDSON RIVER FROM NEW YORK TO ALBANY, 18th century
Scrimshawed powder-horn
Collection of the Geography and Map Division, Library of Congress, Washington, DC

MAP OF THE LAND ALONG FISH-CREEK BELONGING TO THE STATE OF NEW YORK, 1809
39 x 18
Collection of New York State Archives, Albany

MAP OF THE NORTHERN PARTS OF NEW YORK, 1758
28.75 x 33.25
Collection of the Geography and Map Division, Library of Congress, Washington, DC

PILZ ANATOMICAL MANIKIN, c. 1900
Colored lithograph with folding flaps mounted on pasteboard
65 x 58 opened
Collection of the Health Sciences Library, University of North Carolina at Chapel Hill

MOLECULAR GENETICS LABORATORY WORKBENCH, c. 1976–2001
Including the following:
Celera Human Genome Map, 2001;
Leonardo da Vinci drawing and map of gene expression throughout the human body; *Map of Bacterial Genomes;*
Microarray Anaylis in Progress, data from Dr. Joanne Conover, Jackson Laboratory, and Rachel Allen, class of 2003; *Yeast Protein Interactions*, contributed by Dr. Leroy Hood, Director, Institute of Systems Biology, Seattle, Washington
Scanner FMBIO II Multi-View, c. 1995
Courtesy of the New York State Police Forensic Investigation Center

PHRENOLOGY HEAD, c. 1850–1900
Plaster with paper
5.75 x 2.75 x 3.5
Collection of The New York State Museum, Albany

WALL CHART: CORN GENETICS, 1941
Biology Department, Skidmore College

WELLS' NEW DESCRIPTIVE CHART, 1787
Collection of The New York State Museum, Albany

ACKNOWLEDGEMENTS

The World According to the Newest and Most Exact Observations is the first faculty curated exhibition at the Tang Teaching Museum. Collaboration is at the heart of the museum's mission and we are grateful to the many faculty, students, staff and community members who worked to make this ambitious project successful. We would especially like to thank our co-curators and catalogue contributors Bernard Possidente and Richard Wilkinson. Their consistent involvement at every step of the way made the project a truly interdisciplinary experience for all of us.

We are very fortunate to have the opportunity to bring together artwork and maps from a wide variety or sources. Our sincere thanks to the following individuals and institutions for their assistance: Art in General; Seymour Benzer, Boswell Professor of Neuroscience, California Institute of Technology; Ron Burch, New York State Museum; Elizabeth Burke and Abby Messitte, Clementine Gallery; Dr. Joanne Conover, Jackson Laboratory; Kevin Cook, Bloomington Indiana Drosophila Stock Center; Paula Cooper Gallery; Nick Debs, Debs and Co. Gallery; Duke University Medical Center Library; David Egger; Finesilver Gallery; Dr. James Folts, Director of Reference Services, and Richard Andress, Archivist, New York State Archives; Ira Glass, This American Life; Dr. John Hebért, Chief, Ronald Grimm, Specialist in Cartographic History, Margaret Brown, Registrar, and Katie Ryan, Cartographic Specialist for Digital Programs, Geography and Map Division, Library of Congress; Pamela Clapp Hinkle, Director of Communications, Woods Hole Marine Biological Laboratory; Claire Fraser, President, Aimee Turner, Chief Financial Officer, and Lisa McDonald, Technical Training Manager, The Institute for Genomic Research; George Grills, Director, and Ari Hobfoll, Albert Einstein University DNA Chip Microarray Facility, Dr. Leroy Hood, Director, Tawney Biddulph, Assistant to the Director, Institute of Systems Biology, Seattle; Field Horne; Alexis Hubsman, RARE Gallery; Shelley Irwin, Associate Archivist, California Institute of Technology; Carol Jenkins and Diane McKenzie, Health Sciences Library at the University of North Carolina at Chapel Hill; Jimapco; Max and Anne Kaufman; Wendy Durst Kreeger '69 and Douglas A. Kreeger; Heidi Lange, DC Moore Gallery; Terry Lautz, The Henry Luce Foundation; Ricki Lewis; Miriam Mandelbaum, Curator and Arlene Shaner, Reference Librarian, Anne Hillam and Susan Martin, Conservators, Malloch Rare Book Room, New York Academy of Medicine; John Matese, Curator, Stanford Microarray Database; Paul Mercer, Senior Librarian, Manuscripts and Special Collections, New York State Library; Greg Merrill, HT Medical Systems; Roger Milkman and Jerry Kollros, University of Iowa Department of Biology; Rue Moore, Albany Medical College Archives; James Parillo, Executive Director, Doris Lamont, Archivist, Historical Society of Saratoga Springs; Robert Pfannebecker;

Barbara Pilliod, MapInfo Corporation; Andrea Rosen Gallery; Jack Shainman, Jack Shainman Gallery; Jack Somer; Sonnabend Gallery; Sperone Westwater Gallery; Steve Swinton, Supervisor of DNA Services, New York State Police Forensic Investigation Center; Dr. James Watson and Danielle Kovacs, Cold Spring Harbor Laboratory; Darrell Welch, Associate Librarian, New York State Library; Eleanor Williams; Marie Williams, Nature Magazine; Women's Studio Workshop, and an anonymous private collector.

Our special thanks to Alan Voorhees whose generous access to his unique and important collection of atlases and maps not only formed the core of our checklist but helped us frame the content of our project. We are grateful for his family's continued support of the Tang Teaching Museum and its programs. The catalogue is made possible through support from the Henry Luce Foundation and the Friends of Tang. The exhibition has been funded in part by MapInfo Corporation and the Friends of Tang.

Barbara Glauber and Beverly Joel of Heavy Meta built upon the exhibition's themes while designing this exciting catalogue with beautiful photographs by Arthur Evans, Mimmo Capone, and Jon and Anne Abbott. Every member of the Tang staff helped with the project, and we offer special thanks to Brian Caverly, Susi Kerr, Gayle King, Chris Kobuskie, Gavin McKeirnan, Jennifer O'Shea, Barbara Rhoades and Barbara Schrade. Thanks to our installation crew who quickly built a complicated exhibition with grace, Jed Cleary, Shaw Fici, Abraham Ferraro, Steve Martonis, and Thaddeus Smith. We had fabrication help from Ward Fellows, and Skip Grant of Grant Graphics. Also thanks to students Rachel Allen, Daniel Byers, Cindy Fifield, Ariel Magnes, Eric Patridge, and Lauren Restrepo for their assistance. Many faculty members contributed to the exhibition, thanks to Margo Mensing and Kate Leavitt, Art; David Karp, Sociology, Anthropolgy and Social Work; Greg Goodwin, Psychology; Ruth Copans and Amy Syrell, Lucy Scribner Library; Terry Diggory, English; Pat Fehling, Debra Nichols, Denise Smith, and Jeff Segrave, Exercise Science; Janis Ritorto, Chemistry; Sue Ciani, Chris Eddy, Corey Freeman-Gallant, Roy Meyers, Phil Ortiz, Loretta Parsons, Monica Raveret-Richter, Wayne Richter, and Elaine Rubenstein, Biology. Thank you to Robert Kimmerle and Barbara Melville of College Relations, John Danison, Hunt Conard and John Sanders of the Center for Information Technology, and Michael Casey, Tracy Barlok, Barry Pritzker and Paul Hanks of the Advancement office for their support.

Special thanks to Charles Stainback, Director of the Tang for proposing this project and encouraging and supporting us to follow our dreams for the perfect exhibition. We dedicate this catalogue to David Porter and Phyllis Roth's vision for the Tang Museum.

SUSAN BENDER and IAN BERRY

Curators

This catalogue accompanies the exhibition ***The World According to the Newest and Most Exact Observations: Mapping art and Science*** March 3–June 3, 2001

The Tang Teaching Museum and Art Gallery
SKIDMORE COLLEGE
815 North Broadway
Saratoga Springs, New York 12866
T 518 580 8080
F 518 580 5069
www.skidmore.edu/tang

ISBN # 0-9708790-1-6

Library of Congress control number: 2001095944

Available through
D.A.P./Distributed Art Publishers
155 Sixth Avenue, 2nd Floor
New York, New York, 10013
T 212 627 1999
F 212 627 9484

The Tang Teaching Museum would like to thank the artists, copyright holders and rights holders of all images in this exhibition for granting permission to reproduce their works in this publication. Every effort has been made to contact copyright holders prior to publication.

FRONT COVER
NINA KATCHADOURIAN
Handheld Subway (detail),
1996

BACK COVER
DAVID H. BURR|DISTURNELL
Hudson River & Vicinity,
1835

W.H. THOMPSON
Arterial System in Man

Photographs by Arthur Evans, Mimmo Capone, and Jon and Anne Abbott

Designed by Barbara Glauber and Beverly Joel/Heavy Meta
Printed in Germany by Cantz